T0213284

Lecture Notes in Electrical Engineering

Volume 734

The book series *Lecture Notes in Electrical Engineering* (LNEE) publishes the latest developments in Electrical Engineering - quickly, informally and in high quality. While original research reported in proceedings and monographs has traditionally formed the core of LNEE, we also encourage authors to submit books devoted to supporting student education and professional training in the various fields and applications areas of electrical engineering. The series cover classical and emerging topics concerning:

- Communication Engineering, Information Theory and Networks
- Electronics Engineering and Microelectronics
- Signal, Image and Speech Processing
- Wireless and Mobile Communication
- Circuits and Systems
- Energy Systems, Power Electronics and Electrical Machines
- Electro-optical Engineering
- Instrumentation Engineering
- Avionics Engineering
- Control Systems
- Internet-of-Things and Cybersecurity
- Biomedical Devices, MEMS and NEMS

For general information about this book series, comments or suggestions, please contact leontina.dicecco@springer.com.

To submit a proposal or request further information, please contact the Publishing Editor in your country:

China

Jasmine Dou, Editor (jasmine.dou@springer.com)

India, Japan, Rest of Asia

Swati Meherishi, Editorial Director (Swati.Meherishi@springer.com)

Southeast Asia, Australia, New Zealand

Ramesh Nath Premnath, Editor (ramesh.premnath@springernature.com)

USA, Canada:

Michael Luby, Senior Editor (michael.luby@springer.com)

All other Countries:

Leontina Di Cecco, Senior Editor (leontina.dicecco@springer.com)

** This series is indexed by EI Compendex and Scopus databases. **

More information about this series at http://www.springer.com/series/7818

Nima Dokoohaki · Shatha Jaradat ·
Humberto Jesús Corona Pampín · Reza Shirvany
Editors

Recommender Systems in Fashion and Retail

 Springer

Editors
Nima Dokoohaki
KTH - Royal Institute of Technology
Stockholm, Sweden

Shatha Jaradat
KTH - Royal Institute of Technology
Stockholm, Sweden

Humberto Jesús Corona Pampín
Machine Learning Platform
Booking.com
Amsterdam, The Netherlands

Reza Shirvany
Digital Experience-AI & Builder Platform
Zalando SE
Berlin, Germany

ISSN 1876-1100   ISSN 1876-1119 (electronic)
Lecture Notes in Electrical Engineering
ISBN 978-3-030-66105-2   ISBN 978-3-030-66103-8 (eBook)
https://doi.org/10.1007/978-3-030-66103-8

This Springer imprint is published by the registered company Springer Nature Switzerland AG
The registered company address is: Gewerbestrasse 11, 6330 Cham, Switzerland

Contents

Fashion Understanding

The Importance of Brand Affinity in Luxury Fashion Recommendations

Diogo Goncalves, Liwei Liu, João Sá, Tiago Otto, Ana Magalhães, and Paula Brochado

Abstract Recommender systems in the context of luxury fashion need to have expert domain knowledge to offer the informed experience expected by the customers of this sector. Fashion experts have a strong understanding of the intricacies of the fashion scope. The brands and designers are some of the most important features of this landscape and the affinity between them is not always easy to grasp. This paper proposes an application of state-of-the-art NLP techniques to map the knowledge provided by experts in the form of texts. The outcome was a process to extract brand embeddings which mirror the semantic adjacency between all the brands in a catalogue. To test the utility of such an approach, we conducted extensive offline and online tests which have proven the positive reaction of the customers to the new feature. We applied the embeddings as boosting to a base recommender system and we observed an engagement uplift of up to 10%, and applied the embeddings as a content-based recommender to obtain an engagement uplift of up to 3%. Overall, we are confident of the importance of brand affinity information in recommender systems in the luxury fashion domain.

D. Goncalves (✉) · L. Liu · J. Sá · T. Otto · A. Magalhães · P. Brochado
Farfetch, London, UK
e-mail: diogo.goncalves@farfetch.com

L. Liu
e-mail: liwei.liu@farfetch.com

J. Sá
e-mail: joaomario.sa@farfetch.com

T. Otto
e-mail: tiago.otto@farfetch.com

A. Magalhães
e-mail: ana.magalhaes@farfetch.com

P. Brochado
e-mail: paula.brochado@farfetch.com

© The Author(s), under exclusive license to Springer Nature Switzerland AG 2021
N. Dokoohaki et al. (eds.), *Recommender Systems in Fashion and Retail*,
Lecture Notes in Electrical Engineering 734,
https://doi.org/10.1007/978-3-030-66103-8_1

3

1   Introduction

Across the whole fashion e-commerce sector, customers should experience more and more recommendation systems to be tailored to their needs and fashion tastes [5]. In this luxury fashion context, and particularly at Farfetch, customers expect not only a personalised experience, but also an informed opinion regarding latest trends, new arrivals and fashion understanding. Farfetch is the leading platform for online luxury fashion shopping. We count with the biggest catalogue of luxury items in the World with more than 3 million products and more than 10 thousand brands and high-end designers. Moreover, we sell products worldwide to more than 2 million customers. Our customers expect the highest standards regarding shopping and they usually pursue great experiences in their journeys as clients.

Luxury fashion is a form of art, and that has to be taken into consideration whenever we are recommending a product to a possible customer. Experiences and expectations in luxury fashion might be closer to music and *beaux-arts* than to fast fashion shopping. Designers are artists and the work they do in their brands lead to legions of fans to follow them closely and rejecting the idea of art directors moving between brands [13]. Therefore, it's paramount that the recommender systems in such a domain have the strongest fashion understanding of what a brand means on a global and on a personalised scale.

In this context, pure collaborative filtering (CF) approaches are expected to fail due to the behavior of the luxury customer who wants exclusivity. There are many products in a luxury catalogue that contain very few items in stock, being sold-out after just one purchase. This particularity causes products to no longer be available at the time a CF algorithm identifies them as good recommendations to a user. This scenario leads to the need for content-based hybrids, which leverage the users and items information to deliver the level of personalisation needed by a recommender system in the fashion context. There are many ways to incorporate content in recommender systems to create hybrid solutions. However, the quality of the outcome is very dependent on the quality of the provided information.

Our proposal is to implement a neural network embedding model to extract the fashion experts' knowledge regarding brands and incorporate it in our recommender systems. To train our embedding models, we used a curated dataset composed by brands and products descriptions written by Farfetch fashion experts, as well as highly referenced opinion articles about the brands and designers present in our catalogue.

We conducted an offline experiment to compare the embeddings generated by three well known state-of-the-art algorithms—Word2Vec, FastText and Glove [1, 11, 12]. We found that the most suitable approach to present to our customers in an online setting were the embeddings mapped by the Word2Vec model. Then, we conducted an A/B test to expose the users to the new models. We experimented with two settings: (1) recommend products from similar brands at product listing pages (PLP) following a content-based approach; (2) Boosting the current recommender systems at product detail pages (PDP) with contextual information regarding brand affinity. The results showed that this approach has greatly improved the engagement of the users with our recommendation system, leading to uplifts of up to 10% in our engagement metrics.

The main contributions of this work are (1) enlightening the importance of brand affinity in the success of luxury fashion recommendations; (2) presenting that mapping brands as word embeddings learned from fashion experts texts mirrors the fashion affinity between brands; (3) boosting recommendations with extra signals is a fast and effective way to understand the relevance of a feature in recommender systems.

2   Related Works

Embedding methods have been extensively used with great success in various Natural Language Processing (NLP) tasks [1, 11, 12]. Mikolov et al. [11] proposed Word2Vec, a pair of unsupervised algorithms (Skip-Gram and CBOW) able to learn a representation for a word in a dense vector space, in which vectors representing words semantically similar are closer to each other, whereas vectors for words with semantic differences are projected further away. Joulin et al. [1] have revisited the Skip-Gram method from Word2Vec and presented FastText. This algorithm generated the semantic representations not only of the words in the vocabulary, but also of the letters and symbols composing the words in that Corpora. Finally, Pennington et al. [12] proposed a new approach to learn word embeddings called GloVe (Global Vectors for Word Representation). This method is inspired by the Matrix Factorization of co-occurrence of the words in sentences. The authors claim that this approach mirrors the global semantic meaning of a word better than Word2Vec due to the word-word co-occurrence score across all Corpora.

Regarding the specific context of brand representations modelling, Yang and Cho [14] proposed the Brand2Vec approach. The work is an application of the Paragraph2Vec (or Doc2vec) algorithm proposed by Le and Mikolov [7] where the paragraph ids are the brand ids and the texts are the reviews posed by customers from a marketplace. Although Paragraph2Vec work refers the consistent superiority of PV-DM (Paragraph Vectors - Distributed Memory) over PV-CBOW (Paragraph Vectors - Continuous Bag of Words), the authors of Brand2Vec chose the PV-CBOW approach. We are not considering the paragraph modelling for this work mainly due to the interchangeability between brand representations in the texts. For example, the text written by an expert describing a particular brand or designer can include the relationships to other brands. Hence, if the brands are mapped correctly to a single token, a Word2Vec model should be sufficient to map the brand id embedding. Moreover, PV-CBOW and PV-DM are considerably more expensive computationally when comparing to Word2Vec approaches due to the need for embedding computation for the paragraph's representations.

Other automatic feature extraction techniques have been explored in a myriad of machine learning domains. In particular, on fashion related problems, Marcelino et al. [10] proposed the use of a sequence model based on a Long-Short-Term-Memory

(LSTM) neural network to extract features to power a semantic search engine. Unlike [10], we focused on extracting the fashion concept of what is a brand and not on parsing a general query. On a computer vision setting, one can use a Convolutional Neural Network to extract product related visual features from images and use them as side information to hybrid models to power an automated outfit generation [4, 8], or even, a more straightforward task of retrieving the nearest neighbors over the embedding space as a pure content-based filtering recommender [3]. Finally, those visual features could be employed joining with our embeddings to further improve our recommendation engines.

This work's objective is to map the expert domain knowledge from our fashion experts regarding the understating of brands. Given the high availability of textual data, we found that the approach should be focused on NLP techniques. Therefore, we selected three of the more renown methods to explore the brand embeddings learning—Word2Vec, FastText and Glove [1, 11, 12].

3 Methodology

3.1 Data Collection

The main goal of this work is to create an embedding representation of the brands present in our catalogue which could mirror the fashion understanding of the experts.

For that matter, we collected five sources of text data, written in English language by fashion experts as an attempt to reflect the domain knowledge:

- Product information, accounting for more than 3M products:

  - Short description;
  - Long description;
  - Gender;
  - Category levels.

- Brand descriptions, accounting for more than 10k brands;
- Brand DNA, with top brands attributes annotated by Farfetch fashion experts on a set of 200 most popular brands:

  - Art director;
  - Fashion position of the brand;
  - etc.

- Fashion Taxonomy graph, a work conducted at Farfetch leveraging fashion terms and their relationships at both product and brand level. For example, blue is an attribute to a product, but a brand with several products of the color blue will have a strong edge towards that color and can be a brand attribute. This relationship can be constructed to all the concrete and abstract fashion attributes/terms.

- Fashion articles from well renowned sources such as BoF,[1] referring to brands and designers present on the Farfetch catalogue.

3.2 Data Preparation

All five sources of data had the same six preprocessing steps:

1. The text was normalized, accents and special characters were removed and the whole text converted to lowercase;
2. All brands were mapped to a single token, if a brand name has multiple words, the space between them will be connected, e.g. "Yves Saint-Laurent" maps to "yves_saint_laurent". This operation is key for the success of this implementation. If the brand name is not mapped to a single token it will be impossible to obtain a word embedding referring to a brand name. For the case of "Red Valentino", we would have two word embeddings, one for "Red" and another for "Valentino", even though in this case Red is an acronym for "Romantic Eccentric Dress" and not the colour red.
3. The sentences were tokenized;
4. The stopwords were removed;
5. The resulting tokens were stemmed using Snowball method.
6. Finally, some single tokens were joined into pairs (bigrams) in order to improve the semantic representation of frequent pairs of words.

These transformations were applied to the whole data gathered previously and a dataset of 3,806,894 sentences was obtained—the **sentence** dataset.

Two additional datasets were created from the **sentence** dataset:

- **sentence_keep**: the result from applying a filter to sentence dataset, which removes all sentences that have no term found in the Brand Names, Fashion Taxonomy, or Brand DNA—3,263,292 sentences. The hypothesis to test with this dataset is to understand how different are the embeddings representations of a brand when using only sentences referring it.
- **sentence_keep_syn**: synonym mapping between all the words of **sentence_keep** and fashion synonyms identified by experts in Fashion Taxonomy and Brand DNA data—3,263,292 sentences. The hypothesis to test with this data is that if we reduce sparsity, the resulting brand embeddings will have a better semantic representation.

3.3 Brand Affinity Modelling

Mikolov et al. [11] coined the term "word embeddings" on their seminal work presenting the family of non supervised algorithms called Word2Vec. These algorithms

---

[1] https://www.businessoffashion.com/.

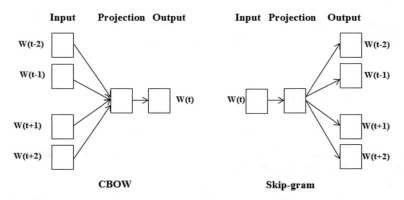

Fig. 1 Word2Vec architectures [11]

aimed at creating probabilistic models with the objective of projecting the whole vocabulary of a Natural Language on the same multi-dimensional space. The training process can be formulated in two ways as observed in Fig. 1:

- **Skip-Gram:** given a word, the model has to be able to predict which words form its semantic context. For example, in any sentence, the model has to predict the neighbor words next to a given word.
- **Continuous Bag-Of-Words (CBOW):** given a context of N words in a sentence, the model has to predict which word is placed in that context.

The authors observed that the semantic representation of the words using Skip-Gram approach was more accurate and at a small additional computational cost when using parallel processing [11]. Therefore, Skip-Gram architecture seems to be the most promising approach to be applied in this work for learning the word embeddings of the fashion experts data.

3.4 Boosting Recommendations with Brand Affinity Information

There are many ways to introduce side information to a recommender system. One can use it as inputs during the training process, but it requires a more complex implementation and resulting models. Another way is relying on ensemble techniques where scores from multiple models are combined into a single final score. In this work, we decided to choose the latter option. The main reason relies on experimenting the importance of a feature before investing heavily in increasing the complexity of the models currently in production.

The objective is to obtain a final score $P(i|c)$ for a given product, i from the catalogue, taking into account the context, c. Context can be any type of information regarding the current navigation intention of the user, such as gender, categories, brands, products, and so on. Hence, $P(i|c)$ can be formulated by the Eq. 1:

$$P(i|c) = (1 - \alpha) \cdot P'_c(i|c) + \alpha \cdot P'_b(i|c) \tag{1}$$

- i: item id;
- c: context;
- α: weight to the brand information;
- P'_c and P'_b are the normalized scores from control and brand recommenders.

where P'_c is the normalized predicted score by our control recommender system; and P'_b is the normalized score given by a content-based recommender using solely the brand embeddings trained by Word2Vec in this project. Both scores are point-wise and related to given product, i, considering the context, c.

Both recommenders are separately trained and α is decided offline before an online test via click propensity optimization using logged data. α is a tunable parameter which represents the strength of the new information to the final recommendations.

4 Offline Experiment

The experiments conducted in this work covered both offline and online settings to fully assess the impact of the different approaches within the recommender system's main objective of escalating our fashion experts' knowledge to our users.

4.1 Offline Setup and Model Selection

In this section, we cover the offline settings, where three algorithms were tested to learn the word embeddings on the aforementioned Corpora—Word2Vec, GloVe, and FastText.

For each algorithm, we tested two different sizes for the embeddings, 120 elements and 300 elements. Considering the three datasets described above (**sentence, sentence_keep** and **sentence_keep_syn**), 18 data-algorithm combinations were tested to understand which approach should be considered to test in a live randomized field trial, in the context of an A/B testing framework.

The offline evaluation of semantic learning is not a straightforward task due to its inherent subjectivity. We decided to follow three different evaluation approaches, one qualitative and two quantitative analysis.

First, we generated pure content-based recommendations using the embeddings obtained by the NLP algorithms by selecting the top 5 nearest neighbors for each brand present in the catalogue. Then, we used two types of pure collaborative recommenders to compare the results with. Note that these recommenders were used solely for the offline experiment.

One of the recommenders was built using navigation data to map the brand-brand relationships from user-product interactions. We built four different recommenders, one for each different source: clicks, add to wishlist, add to bag and orders. We then aggregated the results obtained by each recommender by summing the similarity scores.

The other type of recommender was based on outfit data curated by Farfetch team of stylists. From a pool of 300k outfits, we built a bipartite graph between the brands and the outfit ids to map the co-occurrence of brands in the outfit data which contains expert domain knowledge solely.

For each variation of the NLP algorithms, we computed the offline metrics such as Precision@5, Recall@5 and nDCG@5 between the auxiliary models (navigation and outfits) and the nearest neighbors based on the learned embeddings. For the final model selection, we used the Borda optimal ranking method to aggregate the different sources of results and select a single winner [2].

For subjective analysis, we used the common t-SNE projection of the embeddings to inspect the brands and their neighbors to understand if they make sense regarding the fashion context of the Corpora. This approach is standard practice and can be seen in many works where item embeddings are created [9].

4.2 Results and Discussion of Offline Evaluation

We conducted an offline evaluation of the set of data-algorithm combinations to understand which would be the best approach to implement in a live setting and present it to our users.

Table 1 presents the results for the top 10 best combinations of the models comparing to the navigation based models referred above.

As we can see, the overall results for Precision, Recall and nDCG are very low when comparing the top-5 recommended brands by the fashion experts embeddings to the navigation-based models. Since the latter focus only on the collaborative relationships between users and brands, it seems fair to assume that user interactions derive substantially different results than the recommendations obtained by the fashion experts information.

Nevertheless, the Skip-Gram Word2Vec seems to outperform the competitors regarding all metrics. Regarding the embedding size, it seems that 120 elements tend to outperform larger embedding vectors. At last, the dataset providing better metrics is **sentence** which had no extra steps of preprocessing.

Table 2 presents the results for the top 10 best combinations of the models comparing to the recommendations based on outfit data. Overall, the metrics of Table 2 are higher than those presented in the results of the navigation data (Table 1).

One of the reasons for this to occur might reside in the argument that the brand embeddings proposed in this paper contain expert domain knowledge and the rec-

Table 1 Top 10 offline results comparing to navigation data

Algo.	Size	Dataset	Prec.@5	Rec.@5	nDCG@5	Borda Count
Word2Vec	120	sentence	**0.0069**	**0.0070**	**0.0238**	9
Word2Vec	300	sentence	0.0067	0.0069	0.0229	8
Word2Vec	120	sentence_keep_syn	0.0067	0.0069	0.0232	7
Word2Vec	300	sentence_keep	0.0066	0.0068	0.0228	6
Word2Vec	120	sentence_keep	0.0065	0.0066	0.0232	5
FastText	300	sentence	0.0061	0.0064	0.0212	4
FastText	120	sentence	0.0061	0.0063	0.0211	3
FastText	120	sentence_keep_syn	0.0056	0.0057	0.0190	2
FastText	120	sentence_keep	0.0052	0.0054	0.0180	1
GloVe	120	sentence	0.0038	0.0038	0.0135	0

Table 2 Top 10 Offline results comparing to outfits data

Algo.	Size	Dataset	Prec.@5	Rec.@5	nDCG@5	Borda Count
Word2Vec	120	sentence_keep_syn	**0.0101**	**0.0125**	**0.0339**	9
Word2Vec	120	sentence	0.0094	0.0117	0.0317	8
Word2Vec	120	sentence_keep	0.0090	0.0110	0.0307	7
Word2Vec	300	sentence_keep	0.0087	0.0108	0.0307	6
FastText	120	sentence_keep_syn	0.0086	0.0104	0.0285	5
Word2Vec	300	sentence	0.0086	0.0107	0.0304	4
FastText	120	sentence	0.0085	0.0107	0.0290	3
FastText	300	sentence	0.0083	0.0107	0.0279	2
FastText	120	sentence_keep	0.0079	0.0096	0.0270	1
GloVe	120	sentence	0.0046	0.0054	0.0154	0

ommendation based on the outfit model too. Hence, it's expected that the neighbors found in both settings are more alike. Similarly to the navigation data results, the Skip-Gram Word2Vec with an embedding size of 120 seems to outperform the competitors regarding all metrics. Regarding the dataset preprocessing, **sentence_keep_syn** generates embeddings closer to the results provided by the outfit data.

Regarding the subjective analysis of the quality of the embeddings, we had projected a 2D t-SNE so we could present in this paper for reference. We plotted generic fashion terms and brands in the same space to understand if the terms and the brands would make sense from a fashion point of view (with the help of fashion experts). For the interest of clarity, we're presenting only the embeddings projections of Word2Vec with 120 components, using the **sentence** dataset.

Figure 2 shows the brands which are neighbors to the term "cartoon". As we can observe, the nearest brands are mostly related to kids' clothing, such as Monnalisa.

Fig. 2 t-SNE 2D projection of word embeddings emphasizing the term "cartoon"

Fig. 3 t-SNE 2D projection of word embeddings emphasizing the term "gothic"

Another example presented on Fig. 3 is emphasizing the term "gothic". It's clear that the closest brand to this term is Alexander McQueen. This designer is widely known by the usage of skulls' representations in his designs.

This sort of qualitative analysis is very useful to have a general overview of how much sense the semantic representation of the words make. Overall, the embedding representations make sense. Moreover, our in-house fashion experts tended to agree

Table 3 Model selection via Borda count method

Algo.	Size	Dataset	Borda nav.	Borda outfits	Borda final
Word2Vec	120	sentence	9	8	17
Word2Vec	120	sentence_keep_syn	7	9	16
Word2Vec	120	sentence_keep	5	7	12
Word2Vec	300	sentence_keep	6	6	12
Word2Vec	300	sentence	8	4	12

on the neighbors found for a set of the brands, but, unfortunately, we don't have sufficient survey data to backup their votes and present in this paper.

The decision making process to define which model should we invest in an online experiment has considered the Borda count ranking method. When summing the Borda scores for each variation and offline experiment, we obtain the results presented in the following Table 3.

The final *Borda count* score selects as the best candidate the solution of Skip-Gram Word2Vec with an embedding of 120 elements, using the **sentence** dataset. Both first and second candidates of the final rank seem reasonable for experimenting in a live setting. Nevertheless, to support the choice of the first alternative, we present two additional arguments:

1. We can observe a small difference regarding the offline metrics, but it does not justify the extra complexity of the ETL for processing **sentence_keep_syn**.
2. The productization costs are considerably lower when annotated data, like synonyms, is not necessary.

In conclusion, we productized the following approach:

- **Algorithm**: Skip-gram Word2Vec with an embedding of 120 elements;
- **ETL**: the process to generate the **sentence** dataset, performing the transformations referred in the *Data Preparation* (Sect. 3.2).
- **Content-Based recommender**: using as features the embeddings obtained by the training of the Word2Vec algorithm.

5 Online Experiment

A proper assessment of the impact of different approaches of a recommender system requires a variety of evaluation vectors, from objective to subjective aspects, considering user recommendation interfaces and last but not least, the ultimate intention of the user that can be affected by external factors [6]. For a thorough evaluation of the chosen final algorithm and its impact—Skip-Gram Word2Vec with an embedding of 120 elements—the online experiment was carried out in the form of four randomized field trials in a live environment. The A/B framework chosen focused on all the users reaching the https://farfetch.com portal which were then assigned randomly (probability of 0.5) to the control or alternative groups of each of the experiments.

5.1 Online Setup

The online experiment was composed of four independent streams, as to allow a fair estimation of the algorithms' fitness to fulfill the myriad of touch-points, channels and user interaction points within its journey.

First, the resulting algorithm was interpreted as a content-based recommender on a specific set of listing pages, recommending related brands to the very specific use case of brand listings with very few items. This would be our most aggressive setting as the user's expectations were already frustrated and a successful algorithm would reconvert the user back into continuing the navigation. The null hypothesis, H_0, for this use case was then "the users are not prone to explore similar brands once their expectations have been thwarted".

Next, the algorithm was tested as a boost applied to the current product recommender systems in the following scenarios: two different types of product detail pages with the same null hypothesis, H_0, where "users are equally engaging with the control group recommendations and with recommendations that are enriched with brand affinity data".

Lastly, an edge case for our recommendation system was tested also considering the brand embeddings as a boost for the control recommender system in the form of a operational email. In this case, the user had already purchased and the goal would be to establish the fashion authority by suggesting products from related brands. The H_0 for this use case states that "the users are not susceptible to brand similarity after the purchase".

All the impressions and interactions with the recommendations carousel (as depicted in Fig. 4) are recorded and a comparison of the predetermined engagement metrics[2] dictated, blindly, which alternatives could be *productized*. However, the outcomes of the four streams of online testing reflected a strong engagement gain from the users to this new source of information, across the board.

Fig. 4 Example of the recommendations carousel

---

[2]We reserve the right to not share the metrics in detail due to legal protection.

Table 4 Summary of the online experiments conducted

Recommendations approach	User phase space	Touch-point
Content-based	Consideration	Low-stock listing pages
Brand Affinity Boost	Consideration	Product detail page with stock
Brand Affinity Boost	Consideration	Product detail page without stock
Brand Affinity Boost	Post-purchase	Operational Email

5.2   Results and Discussion of Online Evaluation

As mentioned previously, the online experiments were executed from, essentially, two perspectives on the same brand embeddings: content-based recommendations and brand affinity boost applied on product recommenders. In Table 4, a summary is presented for each of the test settings, which contributed to a full impact analysis on all aspects of the recommender system.

5.2.1   Content-Based Recommendations

The A/B testing framework was configured so that an even split of 50–50% of random visitors would see alternative A, the control, with no recommendations of related brand and, on alternative B, products from the top two adjacent brands we recommended at the bottom of the low-stock listing page.

Figure 5 shows the PDF of the binomial derived from the logged data (impressions and conversions) of the A/B testing outcome on the low-stock PLP page. The

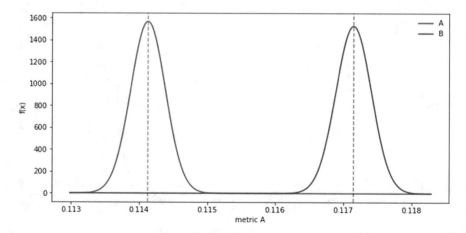

Fig. 5 A/B test results of content-based recommendations

engagement metric presented is not a click-based metric of recommendations since the control group has no recommendations to be clicked. Hence, the null hypothesis can be rejected with a p-value of 1.15×10^{-16}. The distribution of the differences between alternatives expects an engagement uplift between 1.8and 3%, considering the confidence interval of 95%.

These results proved to be very robust in making use of, and enhancing the, subjective relationships between brands on the luxury fashion world within a context of a user that actively looked for a specific brand and was dissatisfied. Such results indicate that fashion-savvy users recognized the validity of the affinity between brands as given by the chosen algorithm to test.

5.2.2   Brand Affinity Boost Recommendations

Using the brand embeddings as a brand affinity boost was implemented in three experiments from two distinct user phases: consideration (two experiments) and post-purchase (one experiment).

From the consideration phase, the two tests were quite similar even though they represented opposite user experiences, mainly at a layout and design level, given the two product detail pages are quite different as they represent distinct states of the product. On both, however, the same overall testing strategy was used: alternative A, the control, represented the current recommendation strategy without the boost from this new brand affinity information. On alternative B, the brand affinity boost was applied to current recommendation strategy, which was the same base strategy as alternative A.

Figure 6 shows the PDF of the binomial distribution derived from the logged data (impressions and conversions) of the A/B testing outcome on the PDP page. The

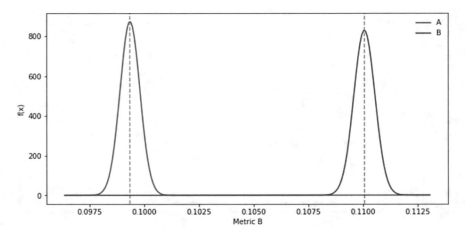

Fig. 6 A/B test results of brand affinity boosted recommendations

metric presented in this A/B test was more quality of recommendations oriented than "Metric A", since the control group was also showing recommendations with the exact same recommendations carousel layout. The null hypothesis can be rejected with a p-value of 0.9×10^{-60}. The distribution of the differences between alternatives expects an engagement uplift between 7.4 and 10.6% for the product detail page without stock experiment, and for the product detail page with stock experiment the same engagement metric expected uplift was between 5.1 and 6.1%, both considering the confidence interval of 95%.

Finally, the post-purchase phase consisted of another experiment executed via email. This test has non-standard configurations and the results are not, in nature, as detailed as the traffic split of a website or device in a live environment. The experiment configuration consisted in an alternative A with a version of an algorithm that showcased only products of the same brand as an input product, whereas alternative B applied the brand affinity boost to another algorithm that did not promote same brand products. In fact, the test here allowed for a direct comparison between same brand vs related brands impact. The engagement metric revealed that the algorithm that used the brand affinity boost, alternative B, was able to outperform alternative A by approximately 15%.

The three brand affinity boost approaches tackled different combinations of user experience and user's expectation. At a product detail page level, the attention of the user is lower as it is in full exploration mode, whereas at the post-purchase phase the expectation was already fulfilled and therefore the intention to engage again is at its lowest. However, in all the three settings, the results showed the user was actively interested in the products that were of a related brand—to note this behaviour towards brands' importance has not been observed in other product attributes of previous experiments. Brand affinity as implemented in this work, proved to be successful on all randomized tests carried out to date.

6 Conclusion and Future Work

In this paper, we presented an effective way of extracting and using brand embeddings, using it as side information to complement more complex recommender systems with the fashion authority expected in the luxury fashion context. The online results have shown a great acceptance from the users exposed to this information. In all the A/B tests performed, the alternatives using the brand affinity information always won against control. The main takeaway was the understanding of brand affinity in the improvement of fashion recommender systems, in the particular context of luxury fashion.

The offline results helped to decide which approach should we choose to take to an online test. However, these decisions are often counterfactual and we cannot derive how well the other approaches would perform in a straightforward manner. Even more evident, it's when the new feature being implemented forces the re-rank of a base recommender updated regularly which drastically changes the outcome.

We find it hard to foresee the outcome of an online test when the new recommender is considerably different than the control. As next steps, we plan to conduct an offline counterfactual evaluation to understand if other NLP approaches would have performed better and run the necessary online experiments to solidify our understanding. Moreover, we have A/B tests ready to start with different variations of the NLP models against the models derived for offline evaluation to understand the relationships between offline and online metrics.

As other future work, we plan as well to incorporate the brand information extracted by these embeddings in our hybrid recommender systems and other recommendation tasks such as outfits generation. We want to conduct a thorough study regarding ensemble optimization and ways to incorporate different sources of information to power recommendations in a straightforward and robust way without exploding in complexity. Finally, we plan to improve the personalization of brand affinity by considering more user navigation signals to improve the context mapping.

References

1. Bojanowski P, Grave E, Joulin A, Mikolov T (2017) Enriching word vectors with subword information. Trans Associat Comput Ling 5:135–146
2. Emerson P (2013) The original borda count and partial voting. Social Choice and Welfare 40. https://doi.org/10.1007/s00355-011-0603-9
3. Gomes J (2017) Boosting recommender systems with deep learning. In: Proceedings of the eleventh ACM conference on recommender systems, RecSys '17, pp 344–344. ACM, New York, NY, USA (2017). https://doi.org/10.1145/3109859.3109926
4. Han X, Wu Z, Jiang Y, Davis LS (2017) Learning fashion compatibility with bidirectional lstms. CoRR abs/1707.05691 (2017). http://arxiv.org/abs/1707.05691
5. Jaradat S, Dokoohaki N, Matskin M (2020) Outfit2vec: Incorporating clothing hierarchical metadata into outfits' recommendation. In: To appear in special issue (fashion recommender systems) in LNSN Springer (2020)
6. Knijnenburg B, Willemsen M (2015) Evaluating recommender systems with user experiments, 2nd edn, pp 309–352. Springer, Germany (2015). https://doi.org/10.1007/978-1-4899-7637-6_9
7. Le Q, Mikolov T (2014) Distributed representations of sentences and documents. pp 1188–1196. PMLR, Bejing, China (2014). http://proceedings.mlr.press/v32/le14.html
8. Li Y, Cao L, Zhu J, Luo J (2016) Mining fashion outfit composition using an end-to-end deep learning approach on set data. CoRR abs/1608.03016 (2016). http://arxiv.org/abs/1608.03016
9. van der Maaten L, Hinton GE (2008) Visualizing data using t-sne
10. Marcelino J, Faria J, Baía L, Sousa RG (2018) A hierarchical deep learning natural language parser for fashion. CoRR abs/1806.09511 (2018). http://dblp.uni-trier.de/db/journals/corr/corr1806.html#abs-1806-09511
11. Mikolov T, Chen K, Corrado G, Dean J (2013) Efficient estimation of word representations in vector space. CoRR abs/1301.3781 (2013). http://dblp.uni-trier.de/db/journals/corr/corr1301.html#abs-1301-3781
12. Pennington J, Socher R, Manning CD (2014) Glove: Global vectors for word representation. In: Empirical methods in natural language processing (EMNLP), pp 1532–1543 (2014). http://www.aclweb.org/anthology/D14-1162

13. team B (2017) The 900 million dollar old celine opportunity. https://www.businessoffashion.com/articles/professional/the-900-million-old-celine-opportunity
14. Yang H, Cho S (2015) Understanding brands with visualization and keywords from ewom using distributed representation. http://dm.snu.ac.kr/static/docs/TR/SNUDM-TR-2015-07.pdf

Probabilistic Color Modelling of Clothing Items

Mohammed Al-Rawi and Joeran Beel

Abstract Color modelling and extraction is an important topic in fashion. It can help build a wide range of applications, for example, recommender systems, color-based retrieval, fashion design, etc. We aim to develop and test models that can extract the dominant colors of clothing and accessory items. The approach we propose has three stages: (1) Mask-RCNN to segment the clothing items, (2) cluster the colors into a predefined number of groups, and (3) combine the detected colors based on the hue scores and the probability of each score. We use Clothing Co-Parsing and ModaNet datasets for evaluation. We also scrape fashion images from the WWW and use our models to discover the fashion color trend. Subjectively, we were able to extract colors even when clothing items have multiple colors. Moreover, we are able to extract colors along with the probability of them appearing in clothes. The method can provide the color baseline drive for more advanced fashion systems.

Keywords Color clustering · Deep learning · Clothing · Fashion trends · k-Means · Gaussian mixture model

1 Introduction

Color modeling and extraction (Fig. 1) has long been an important topic in many areas of science, business and industry. Color is also one of the fundamental components of image understanding. Due to problems of color degradation over time and the possibility of having a large number of colors in every image, this area is still under extensive research [9, 18].

Color is one of the important cues that attracts customers when it comes to fashion. Fashion designers and retailers understand this and they usually make use of color services, such as the catalogs provided by Pantone (pantone.com). Moreover, fashion

M. Al-Rawi (✉) · J. Beel
ADAPT Centre, Trinity College Dublin, Dublin, Ireland
e-mail: alrawim@tcd.ie

J. Beel
e-mail: joeran.beel@tcd.ie

© The Author(s), under exclusive license to Springer Nature Switzerland AG 2021
N. Dokoohaki et al. (eds.), *Recommender Systems in Fashion and Retail*,
Lecture Notes in Electrical Engineering 734,
https://doi.org/10.1007/978-3-030-66103-8_2

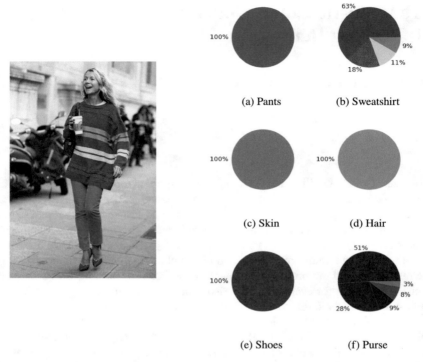

(a) Pants   (b) Sweatshirt

(c) Skin   (d) Hair

(e) Shoes   (f) Purse

Fig. 1 Colors and their possibilities are extracted from the image shown to the left. Sweatshirt colors' names are: 'Egyptian blue', 'Rosso corsa', 'Old gold', and 'Spanish viridian'; pants color name is 'USAFA blue'. Best viewed in color

and AI have found a common ground in recent years. In this regard, companies race to build products and services to serve customers. The impact of deep learning and the development of other AI methods provide companies with unprecedented means to achieve their goals. Amazon and StitchFix are providing their customers with the so called "Personal Styling Service," which is semi-assisted by fashion stylists [7, 10]. Facebook is building a universal product understanding system where fashion is at its core [3, 4]. Zalando researchers proposed in [22] a model for finding pieces of clothing worn by a person in full-body or half-body images with neutral backgrounds. Some other companies are dedicated to build sizing and fitting services [14, 28, 32]. However, efforts dedicated for explicitly making use of color values in fashion AI are somehow limited. This is because most, if not all, products/works rely on the use of color tags. In addition, deep learning models, which are now the cutting edge technology used for several fashion AI apps, are color agnostic. That is, they do a remarkable prediction job without explicitly extracting the color values. This is because the color feature is implicitly extracted alongside other spacial features at the convolution layer.

Several works treated color as a classification problem. Color classification has been the topic of [27]. Other works used deep learning to predict one of eleven basic color names [29] and one of 28 color names [31]. Clearly, color classification has the disadvantage of dealing with a small number of tagged colors. This is a serious problem because of the failure to keep pace with the millions of colors that the human vision system can perceive [17]. This suggests that color extraction should be treated as a regression and not a classification problem.

There are a few works dedicated to extracting the main color values from images using regression methods. The authors of [23] presented a method for extracting color themes from images using a regression model trained on color themes that were annotated by people. To collect data for their work, the authors asked people to extract themes from a set of 40 images that consisting of 20 paintings and 20 photographs. However, such data-driven approach may suffer from generalization issues because millions of colors exist in the real world. The authors of [33] used k-means algorithm on an input image to generate a palette consisting of a small set of the most representative colors. An iterative palette partitioning based on cluster validation has been proposed in [18] to generate color palettes.

Clustering methods have been proposed to extract colors from images much earlier than classification and regression methods. Automatic palette extraction has been the focus of [12, 13] in which a hue histogram segmentation method has been used. The hue histogram segmentation has a disadvantage not only that it is affected by the saturation and intensity values, but also singularities when the saturation values are zero. Other works, as in [8], pointed out the advantages the fuzzy-c-means can provide over k-means even though they aimed at using it for color image segmentation. The main problem with clustering algorithms is knowing the number of colors a prior. Choosing a low value will result in incorrect color values if the image has more colors than the one used to build the clustering model. On the other hand, using a high number of clusters has the drawback of extracting several colors, some with proximate hue values. This makes it really difficult to extract colors accurately.

In this work we propose a multistage method intended for modeling and extracting color values from the clothing items that appear in images. Our pipeline makes use of Mask-RCNN [15], which is a deep neural network aimed to solve instance segmentation problem, to segment clothing items from each image. Next, we cluster the colors into a large number of groups; and then, merge the resultant colors according to their hue and probability values. We use a probabilistic model because it is possible for clustering algorithms to yield more than one color value for each pure color.

2   Color Modelling and Extraction

Stored as images, the colors of clothes usually suffer from severe distortions. The life of a clothing item, color printing quality, imaging geometry, amount of illumination, and even imaging devices affect how colors appear in images. Additionally, clothing items differ in material, fabric, texture, print/paint technology, and patterns. Below, we introduce the mathematical intuition behind the color distribution, followed by a method for extracting colors from clothing items.

2.1 Mathematical Modelling

Without regard to complexity and color distortions, the single color and single channel image should have a Dirac delta density distribution defined as follows:

$$\delta(x|\mu, \sigma) = \lim_{\sigma \to 0} \frac{1}{\sqrt{\pi}|\sigma|} \exp\left(\frac{-(x-\mu)^2}{\sigma^2}\right), \tag{1}$$

where μ is the intensity value. RGB color images have three channels; hence, we write the density distribution of a single-color RGB image is given by:

$$\Delta(\mathbf{x}|\boldsymbol{\mu}, \boldsymbol{\Sigma}) = \lim_{\Sigma \to 0} \frac{1}{\sqrt{\pi|\Sigma|}} \exp\left(-(\mathbf{x}-\boldsymbol{\mu})^T \Sigma^{-1}(\mathbf{x}-\boldsymbol{\mu})\right), \tag{2}$$

where Σ is a $q \times q$ sized covariance matrix, q is the number of channels ($q = 3$ for an RGB image), $\boldsymbol{\mu}$ and $\mathbf{x}$ are q sized vectors, and $|\Sigma|$ is the determinant of Σ. The distribution in (2) is a point in 3D space defined by the value $\boldsymbol{\mu}$. In the real world, and even when the imaging setting is typical, the color distribution of each channel will be normally distributed around μ. This can be denoted for the one channel single-color case as follows:

$$n(x|\mu, \sigma) = \frac{1}{\sqrt{2\pi}\sigma} \exp\left(\frac{-(x-\mu)^2}{2\sigma^2}\right), \tag{3}$$

where σ parameter relates to color dispersion. For a single-color RGB image, the density is a multivariate Gaussian distribution given by:

$$\mathcal{N}(\mathbf{x}|\boldsymbol{\mu}, \boldsymbol{\Sigma}) = \frac{1}{\sqrt{2\pi|\Sigma|}} \exp\left(-(\mathbf{x}-\boldsymbol{\mu})^T \Sigma^{-1}(\mathbf{x}-\boldsymbol{\mu})\right). \tag{4}$$

And for a multicolor image, we model the colors as a Gaussian mixture model prior distribution on the vector of estimates, which is given by:

$$p(\mathbf{x}) = \sum_{i=1}^{K} \phi_i \mathcal{N}(\mathbf{x}|\boldsymbol{\mu}_i, \boldsymbol{\Sigma}_\mathbf{i}), \tag{5}$$

where

$$\mathcal{N}(\mathbf{x}|\boldsymbol{\mu}_i, \Sigma_\mathbf{i}) = \frac{1}{\sqrt{(2\pi)^K |\Sigma_\mathbf{i}|}} \exp\left(-(\mathbf{x} - \boldsymbol{\mu}_i)^\mathrm{T} \Sigma_\mathbf{i}^{-1}(\mathbf{x} - \boldsymbol{\mu}_i)\right), \qquad (6)$$

K denotes the number of colors, and the i_{th} vector component is characterized by normal distributions with weight ϕ_i, mean $\boldsymbol{\mu}_i$ and covariance matrix Σ_i.

Expressed as a Gaussian mixture model, the color distribution also becomes extremely complicated when the clothing item has more than one color. However, the distribution may become more complicated because there is a possibility of frequency deviation of the colors during the imaging process. In this case, the Gaussian mixture model prior distribution on the vector of estimates will be given by:

$$p(\mathbf{x}) = \sum_{i=1}^{K'} \phi_i \mathcal{N}(\mathbf{x}|\boldsymbol{\mu}_i, \Sigma_\mathbf{i})), \qquad (7)$$

where K' denotes the total number of colors or model components such that $K' = K + K_f$, and K_f denotes the number of new (fake) colors generated during image acquisition. In fact, even the human vision system can perceive extra/fake colors due to the imaging conditions. One example is the white reflection we perceive when looking at black leather items. Another problem that perturbs the Gaussian mixture model is the use of 8 bits per pixel by both imaging devices and computers. This 8-bit representation for each channel results in truncating the pixel values close to the lower-bound and/or upper-bound, *i.e.* values close to 0 or 255 for an 8-bit per pixel image. This indicates that there will always be some incorrect color distributions in real-world images. This suggests that the estimation of K' has to always be heuristic.

It would be useful if we can adopt the model we derived in (7) to extract colors from clothing items. Although unsupervised clustering from Gaussian mixture models of (7) can be learned using Bayes' theorem, it is difficult to estimate the Gaussian mixture model of the colors without knowing the colors' ground-truth and the value of K'. Moreover, these models are usually trapped in local minima [16]. Luckily, it has been shown in [26] that k-means clustering can be used to approximate a Gaussian mixture model. We are going to make use of k-means as part of our multistage method, and the full pipeline is illustrated next.

2.2 Clothing Instance Segmentation

We train a Mask-RCNN model [15] to segment all clothing items from an input image. Let this procedure be denoted as:

$$S = \mathbb{M}(\mathbf{f}), \qquad (8)$$

where:

- **f** denotes the input image, which is a vector of (triplet) RGB values.
- $\mathbb{M}$ denotes the Mask-RCNN model.
- $S = \{s_0, s_1, ...\}$ is a set of images. Each element in S is a vector that denotes a segmented clothing item. We remove the background from each element in S and we store it as a vector of (triplet) RGB values.

Then we use the trained Mask-RCNN model to segment clothing items at color extraction phase.

2.3 Extracting the Main Colors

After segmenting the clothing items of **f**, we extract the main/dominant colors in each of them. Our color extraction still has a few phases. We first use a clustering algorithm to to obtain the main RGB color components in the clothing segment. Clustering allows us to reduce the number of colors in the image to a limited number. We denote this procedures as follows:

$$\mathbf{c} = \Psi_k(\mathbf{s}), \tag{9}$$

where:

- Ψ_k denotes the clustering model.
- **s** denotes a vector of one clothing item that we obtain via (8); we drop the subscript i from $\mathbf{s}_i$ to simplify notation.
- $\mathbf{c} = [c_0, c_1, ..., c_k]$ is the resultant cluster centers.

Let N_{c_i} be the number of pixels of color c_i and N_s be the total number of pixels in **s**, we estimate the probability of each color as follows:

$$p_i = \frac{N_{c_i}}{N_s}, \tag{10}$$

where $i = 0, 1, ..., k$. Let $\mathbf{p} = [p_0, p_1, ..., p_k]$ be a vector that contains the probability values of each color in **c**.

Clearly, the number of resultant colors equals to the designated number of clusters, k. We use k-means++ clustering algorithm [2, 25] to obtain the main RGB color components in the clothing segment. k-means++ is highly efficient and able to converge in $O(log\ k)$ and almost always attains the optimal results on synthetic datasets [2]. We also use fuzzy c-means clustering [6] as an additional comparison method. Fuzzy c-means is much slower than k-means but it is believed to give better results than the k-means algorithm [6].

The main problem in clustering algorithms is the estimation of the value of k, *i.e.* how many colors are there in the clothing item. Choosing a low value will result in incorrect color values if the image has more colors than k. Choosing a high k value will result in more colors than expected, some of which have the same color but differ in tint/shade. Hence, our approach is to chose a high k value, and then merge the colors according to the hue value and probability of each color. This is illustrated next.

2.4   Merging Pure Colors that Have Different Tints/shades

Our next step is to determine whether or not to merge the resultant colors $\mathbf{c}$ in (9). Towards this end, we use a 1D clustering approach that relies on the variations between different color hues. This allows us to merge the colors based on hues that are similar, if any. We denote the procedure as follows:

$$G = \Phi(\mathbf{h}),\qquad\qquad(11)$$

where:

- Φ denotes a 1D clustering method.
- $\mathbf{h} = [h_0, h_1, ..., h_k]$ is a vector containing the hue components calculated for each color value of $\mathbf{c}$.
- G is a set containing subsets $\{G_0, G_1, ...\}$. Each subset has labels grouped according to the hue values such that: $g_i \in \{0, 1, ..., k\}$ and $G_i \cap G_j = \varnothing$.

To implement Φ, we try two different methods:

(1) Mean-Shift [30]. The Mean-Sift method is feature-space analysis technique that can be used for locating the maxima of a density function. It has found applications in cluster analysis in computer vision and image processing [11, 24].
(2) We also propose a novel 1D clustering algorithm based on differentiating the hue values, and then finding the extreme points according to the maxima of the derivative. We do this as follows:

$$G = \underset{index}{\operatorname{argmax}}\ \mathbf{h}',\qquad\qquad(12)$$

where $\mathbf{h}'$ denotes the derivative of $\mathbf{h}$.

2.5   Probabilistic Modelling

After clustering and color merging, we use the color probabilities given in (10) to estimate the final color. Our suggestion here is driven by the fact that colors with

higher likelihood should dominate the final color. Accordingly, we propose to extract
the colors according to the following probabilistic model:

$$d_j = \underset{i \in G_j}{\mathbb{E}} \,[\mathbf{p} \times \mathbf{c}], \qquad (13)$$

where $j = \{0, 1, ...|G|\}$, $\mathbf{p}$ is given in (10) and $\mathbf{c}$ in (9). The Mean-Shift method also
results in a G set, one can similarly use (13) to estimate the colors. The final colors'
probabilities can then be calculated using:

$$\hat{p}_j = \frac{N_{d_j}}{N_s}, \qquad (14)$$

where N_{d_j} is the number of pixels of color d_j. Hence, our merging algorithm averages
the RGB values according to the category and probability of each hue. As a simple
example illustrating this approach, one can imagine the color produced by mixing
10% dark blue with 90% light blue in oil paints.

2.6   Color Names

We use the CIELAB color space to match a query RGB value to a lookup table of
color names, *i.e.* color names dictionary. It is believed that CIELAB color is designed
to approximate human vision [5, 21]. For a query color value, the best match is the
color with the minimal Euclidean distance in CIELAB color space to a color in a
dictionary of colors. We build the color name dictionary from "Color : universal
language and dictionary of names" [19].

3   Results

We perform a few experiments to investigate the methods we propose. To reduce
clutter, we do not show the color names in most of the figures. We opt instead
to provide the probability associated to each extracted color. As there are no color
ground-truth, the performance is, unfortunately, subjective. The implementation code
can be found in [1].

3.1   The Effect of Number of Clusters

Figure 2 illustrates the effect of only using the k-means compared to our multistage
probabilistic approach. Using $k = 20$, one stage k-means extracts different degrees

(a) Input image

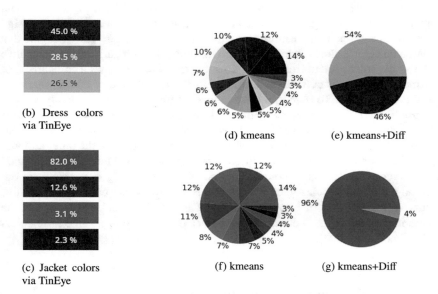

(b) Dress colors via TinEye

(c) Jacket colors via TinEye

(d) kmeans

(e) kmeans+Diff

(f) kmeans

(g) kmeans+Diff

Fig. 2 Use of k-means versus our proposed multistage method 'k-means+Diff'. We use $k = 20$ in all these tests. The dress has three colors; white, black, and gray as shown in (**a**). Our method is able to extract two colors of which 46% black and 54% gray as shown in (**e**). Using only k-means gives 16 shades of gray as shown in (**d**). The jacket has one color (blue), and we are able to extract 96% blue and 4% gray as shown in (**g**). Using only k-means on the Jacket gives 13 shades of blue as shown in (**f**). For comparison, we show the results of using the TinEye Color Extraction Tool in (**b**) and (**c**). Best viewed in color

of blue from the 'jacket' that the lady worn, and different degrees of black and gray from the dress. For the same jacket, our method results 94% blue and 4% gray. For the dress, our method results in 54% light gray and 46% black. In all our models, we remove the color if its probability is less than $1/2k$. We do this to reduce the number of colors based on color probability, *i.e.* as low probability indicates trivial/incorrect color. This is one reason we get lower than k colors when only the k-means is used. Clearly, the multistage clustering we propose helps reduce the colors further as we merge colors of similar hues.

3.2   Comparison with Color Extraction Tools

We compare the colors we extract using our proposed method with some of the available commercial tools. For this purpose, we choose Canva (canva.com) and TinEye MultiColor Engine (tineye.com). Canva is one of the commercial design tools that is equipped with a color extractor. TinEye is an image search engine that is based on color features. It must be noted that our color comparisons are subjective because the color ground-truth values do not exist. Figure 3 shows a comparison of our method with that of TinEye. The input image is a picture of a lady wearing a multi colored dress. To highlight these colors for the reader, we manually marked at least 10 colors in the segmented dress, shown in Fig. 3. Results of using TinEye MultiColor engine does not produce white, dark blue, and other degrees of black. Our proposed methods outperform the TinEye MultiColor Engine as the latter and was not able to extract white color and different degrees of black. Moreover, our method extracts more representative colors than that of TinEye. In Fig. 4, we compare our method with Canva tool. Again, our method is able to extract better representative colors. The upper part of Fig. 4 shows that Canva tool did not extract one of the colors in the t-shirt (the 'Jazzberry jam' color) and also produce some false colors (Silver and Dark Slate).

3.3   Color Distributions of Fashion Data

Investigating the color distribution of fashion is important. It can be used, among other things, to explore fashion trends. We present in Fig. 5 color distributions of selected items of ModaNet. Although each item may have more than one color, we present the distributions of the colors with the highest probability, these denote the dominant colors. Using only colors with the highest probability simplifies the graphs. We color each point, mimicking one item, according to the extracted color. These graphs provide a nice visual representation about fashion items.

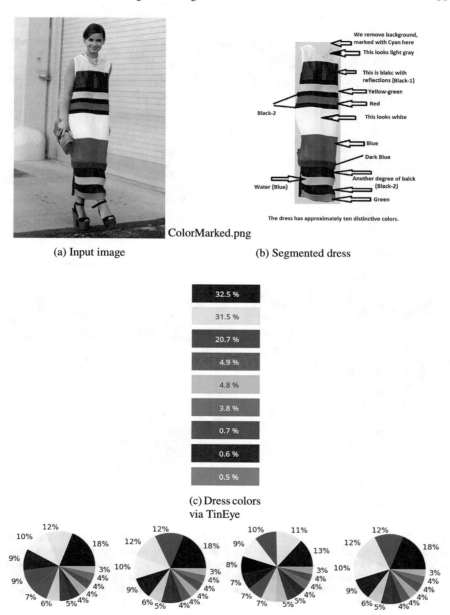

(a) Input image

(b) Segmented dress

(c) Dress colors
via TinEye

(d) kmeans+Diff (e) kmeans+MeanShift (f) FCM+Diff (g) FCM+MeanShift

Fig. 3 Comparison of color extraction methods. We manually marked at least 10 colors in the segmented dress shown in (**b**), Results of a commercial color extraction tool are shown in (**c**). Variants of our proposed method are shown in (**d**), (**e**), (**f**) and (**g**); FCM denotes using Fuzzy-C-Means. Our proposed methods outperformed the Multicolor Engine (tineye.com), as the latter was not able to extract the white color and different degrees of black. Best viewed in color

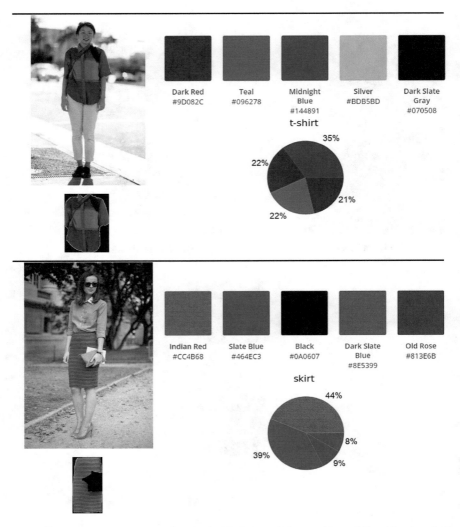

Fig. 4 Comparison of our proposed method with Canva color extraction tool. In each row, colors extracted using our method (lower-right pie-chart) and Canva commercial package (top-right array-chart; www.canva.com). We extract colors from one clothing item segmented out via Mask-RCNN model. To reduce clutter, we did not include the color names, although we can obtain them via the *ColorNames* class. For example, colors of the t-shirt image are: 'Dark cornflower blue', 'Rufous', 'Blue sapphire (Maximum Blue Green)', and 'Jazzberry jam'. Best viewed in color

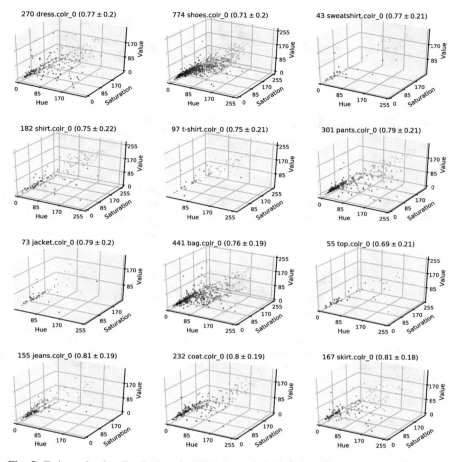

Fig. 5 Estimated color distributions in HSV space. Distributions generated for each item from Clothing Co-Parsing dataset. Titles above each sub-figure are: number of clothing item, item name, *colr_0* is the dominant color (highest probability), followed by mean ± standard-deviation of probabilities. Best viewed in color

3.4 Fashion Color Trend

We make use of the methods we propose to obtain fashion color trend from a set of images. We extract the color trend from Chanel's 2020 Spring-Summer season. The images we use are from https://bit.ly/2WQzKwp. We use Mask-RCNN that we trained with ModaNet dataset to segment clothing items. Each pie-chart denotes one segmented item. We present the results in Figs. 6, 8, and 7.

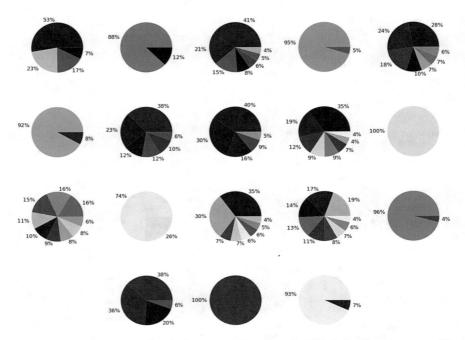

Fig. 6 (Dress) Fashion color trend we extracted from Chanel's 2020 Spring-Summer season. We use Mask-RCNN that we trained with ModaNet dataset to segment Dresses. Each pie-chart denotes one segmented item. Original model images can be reached via https://bit.ly/2WQzKwp. Best viewed in color

4   Discussion and Conclusion

4.1   Gaussian Mixture Model Versus K-Means

Although the k-means method has long been used in color extraction, we wanted to find a link between k-means and the color model we present in (6). It seems that a recent study [26] has found that link. The authors showed that k-means (also known as Lloyd's algorithm) can be obtained as a special case when truncated variational expectation maximization approximations are applied to Gaussian mixture models with isotropic Gaussians. In fact, it is well-known that k-means can be obtained as a limit case of expectation maximization for Gaussian mixture models when $\sigma^2 \to 0$ [26]. Bur according to our color model in (1), $\sigma^2 \to 0$ indicates that the clothing item is imaged in a an idealistic conditions; and that's intractable in reality. Nevertheless, the work of [26] gives some legitimacy and justification for using the k-means as an approximation of Gaussian mixture models.

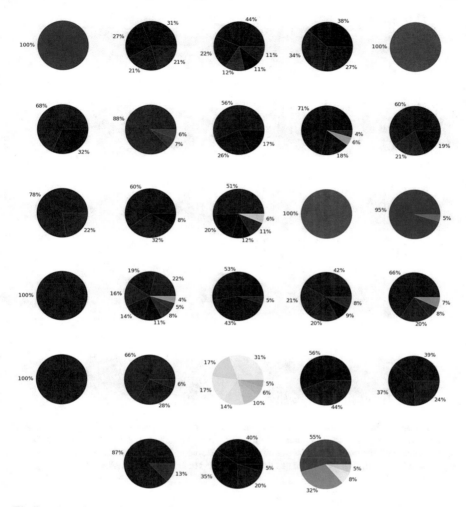

Fig. 7 (Pants) Color fashion trend that we extracted from Chanel's 2020 Spring-Summer season. We use Mask-RCNN that we trained with ModaNet dataset to segment Pants. Each pie-chart denotes one segmented item. Original model images can be reached via https://bit.ly/2WQzKwp. Best viewed in color

4.2 Probabilistic Color Model

One of the biggest problems in color extraction from digital images is dealing with the many colors that a color extractor may reveal. Each clothing item has a set of colors that are, at least, distinctive to the human vision system. And even if we exclude clothes of complex colors and a fractal pattern shapes, the problem remains. One can think of the clustering approach as finding the colors' outcome by averaging

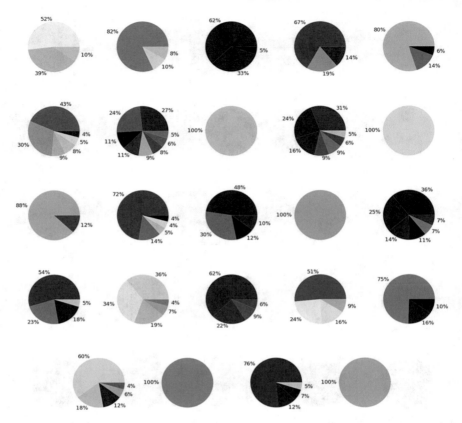

Fig. 8 (Outer) Fashion color trend that we extracted from Chanel's 2020 Spring-Summer season. We use Mask-RCNN that we trained with ModaNet dataset to segment Outer (coat/jacket/suit/blazers/cardigan/sweater/Jumpsuits/ Rompers/vest). Each pie-chart denotes one segmented item. Original model images can be reached via https://bit.ly/2WQzKwp. Best viewed in color

the neighboring color values. However, using a low k value may result in averaging different colors of the clothing item. This may lead to loosing the pure colors, if any. Setting k to a high value may result in more colors, such that some colors have similar hues but different shades and/or tints. This variation is expected due to the imaging conditions, amount of light and shadows. Therefore, using a high k value is a better option because it can generate more pure colors. Then, reducing the number of pure colors based on the hue values and associated probability is one way that we find successful to generate more representative colors. The probabilistic model, $\mathbb{E}[\mathbf{p} \times \mathbf{c}]$, we propose in (13) not only performed well in merging colors of similar hues, but has also a natural intuition. This can be justified by the fact that colors appear in a probabilistic manner and can therefore be extracted in the same way.

4.3 Making Use of Color Extraction in Fashion

It is possible to make use of the extracted colors to build color distributions and trends. A color distribution provides a global view of the colors used in a fashion collection. This can be used across different fashion collections in order to have an idea about the colors in each collection. We can see for example how the shoes and bag distributions have a similar shape, which might reflect the color matching between the two groups.

Color value extraction can be highly beneficial if augmented with other predictions obtained via deep learning. Such augmentation can be used to retrieve clothing items from online stores in the event that the items do not have color tags; Or, when items contain many colors in certain percentages. For example, some customers, or fashion designers, may be interested in searching online stores for an item that has the following colors: 40% navy blue, 30% gold yellow, 20% baby pink, and 10% Baby blue eyes. Such palette can be extracted from an item they have seen; similar to the street2shop use-case paradigm [20]. In addition, this probabilistic color palette can be used to build fashion matching models that can be used as part of recommender systems. Clearly, colors extracted from clothes can find numerous applications in the fashion industry. They can be used to help designers in their work, to retrieve clothing items alongside other attributes predicted by AI models, be part of personal shopping and styling apps, and as a vital component of recommender systems.

4.4 Color Perception and Evaluation

Color is usually stored as triplet values in images identified by the red, green and blue channels. However, obtaining the churn of these triplet values via deep learning is a difficult task, although it may seem simple. This is because the extracted colors should conform to the color perception in human vision. Therefore, regression will be better than prediction models. Furthermore, colors should be provided as palette or theme values and not as a few number of tags; which is not yet the case in several fashion datasets. The major problem is that (1) a model that extracts the exact color values is not available; (2) the colors ground-truth is also not available; and (3) if we build a model that extracts the color values, we do not have the colors ground-truth to verify it. This led us to use a subjective measure to judge colors extracted from clothing items. It must be noted that the use of subjective scales may be problematic due to the wide range of millions of colors and differences of opinion.

4.5 Future Prospects

There are many ways to improve this work. For example, if we know the material or fabric type of the piece of clothing, then we can extract colors based this trait. We can do this using a parametric model that takes into account the color that we want to recover according to a parameter denoting the trait. Returning to the Gaussian model mentioned earlier, we can rewrite (3) as follows:

$$\mathcal{N}(x|c_i, \sigma, \gamma) = \frac{1}{|\gamma\sigma|\sqrt{2\pi}} \exp^{\left(-(x-c_i)/2\gamma\sigma\right)^2}, \qquad (15)$$

where γ is a parameter that one can use to modulate the material type. This is based on the perception that the materials used in the clothes affect the amount of light the camera sensor receives. For example, when leather is the material in use, there will be quite a few reflections camouflaging the real color. Hence, color extraction methods would benefit from γ to either estimate the correct number of colors in the item, or quantify the degree of glittering or shininess. This case would also be interesting if a customer wants to search and retrieve an item with these glittering and shininess traits.

We previously indicated that evaluating color extraction methods is an intractable task unless the data set and its ground-truth are derived from a large number of colors. We aim in our next work to create a large dataset characterised by a large number of colors. Color values are derived from several distributions and are then manipulated to generate true colors. This is done by applying filters that simulate the imaging process. These filters can be generated using image processing tools; or empirically via acquiring printed versions of computer generated source color images. The print quality, imaging device, and many other conditions (*i.e.* indoors or outdoors imaging locations) will affect the acquired images and the computed inverse filter(s). This way we will be able to generate the source, computer generated, and (degraded) ground-truth colors. The dataset can then be used for comprehensive evaluation of our probabilistic color extraction method, as well as other color extraction methods.

Acknowledgements This research was conducted with the financial support of European Union's Horizon 2020 programme under the Marie Sklodowska-Curie Gran Grant Agreement No. 801522 at the ADAPT SFI Research Centre, Trinity College Dublin. The ADAPT SFI Centre for Digital Content Technology is funded by Science Foundation Ireland through the SFI Research Centres Programme and is co-funded under the European Regional Development Fund (ERDF) through Grant No. 13/RC/2106_P2.

References

1. Al-Rawi M (2020) https://github.com/morawi/FashionColor-0
2. Arthur D, Vassilvitskii S (2007) K-means++: The advantages of careful seeding. In: Proceedings of the eighteenth annual ACM-SIAM symposium on discrete algorithms, SODA '07, p.

1027–1035. Society for Industrial and Applied Mathematics, USA (2007)

3. Bell SM, Liu Y, Alsheikh S, Tang Y, Pizzi E, Henning M, Singh KK, Parkhi OM, Borisyuk F (2020) Groknet: Unified computer vision model trunk and embeddings for commerce

4. Berg T, Bell S, Paluri M, Chtcherbatchenko A, Chen H, Ge F, Yin B (2020) Powered by ai: Advancing product understanding and building new shopping experiences. https://tinyurl.com/yyomybc9

5. Best J (2017) Colour Design Theories and Applications. A volume in Woodhead Publishing Series in Textiles. Elsevier Ltd. : Woodhead Publishing, Duxford, United Kingdom

6. Bezdek JC, Ehrlich R, Full W (1984) Fcm: The fuzzy c-means clustering algorithm. Comput. Geosci. 10(2):191–203

7. Biron B (2019) Amazon launched a new personal-styling service that works a lot like stitch fix. https://tinyurl.com/yy86ajhj

8. Capitaine HL, Frèlicot C (2011) A fast fuzzy c-means algorithm for color image segmentation. In: Proceedings of the 7th conference of the European Society for Fuzzy Logic and Technology (EUSFLAT-11), pp 1074–1081. Atlantis Press, Aix-les-Bains, France (2011). https://doi.org/10.2991/eusflat.2011.9

9. Cheng WH, Song S, Chen CY, Hidayati SC, Liu J (2020) Fashion meets computer vision: a survey

10. Co SF (2020) Personal styling for everybody. https://www.stitchfix.com/

11. Comaniciu D, Meer P (2002) Mean shift: a robust approach toward feature space analysis. IEEE Trans Pattern Anal Mach Intell 24(5):603–619

12. Delon J, Desolneux A, Lisani JL, Petro AB (2005) Automatic color palette. In: IEEE International conference on image processing 2005, vol 2, pp II–706. IEEE, Genova, Italy (2005)

13. Delon J, Desolneux A, Lisani JL, Petro AB (2007) Automatic color palette. Inverse Prob Imaging 1(2):265–287

14. EasySize (2020) https://www.easysize.me/

15. He K, Gkioxari G, Dollár P, Girshick, RB (2017) Mask r-cnn. In: Proceedings of the 2017 IEEE international conference on computer vision (ICCV), pp 2980–2988. IEEE, Italy (2017)

16. Jin C, Zhang Y, Balakrishnan S, Wainwright MJ, Jordan MI (2016) Local maxima in the likelihood of gaussian mixture models: Structural results and algorithmic consequences. In: Proceedings of the 30th international conference on neural information processing systems, NIPS'16, p. 4123–4131. Curran Associates Inc., Red Hook, NY, USA (2016)

17. Judd DB, Wyszecki G (1975) Color in business, science and industry. Wiley Series in Pure and Applied Optics (third ed.). New York: Wiley-Interscience, New York, NY (1975)

18. Kang JM, Hwang Y (2018) Hierarchical palette extraction based on local distinctiveness and cluster validation for image recoloring. In: Proceedings of 2018 ICIP, pp 2252–2256

19. Kelly, KL, Judd DB (1976) Color: universal language and dictionary of names. NBS special publication - 440. Department. of Commerce, National Bureau of Standards, Washington, DC: UD

20. Kiapour M, Han X, Lazebnik S, Berg A, Berg T (2015) Where to buy it: Matching street clothing photos in online shops. In: 2015 IEEE international conference on computer vision (ICCV), pp 3343–3351. IEEE Computer Society, Santiago, Chile

21. Kremers J, Baraas RC, Marshall NJ (2016) Human Color Vision. Springer Series in Vision Research, vol 5, 1 edn. Springer International Publishing, Cham, Switzerland (2016)

22. Lasserre J, Bracher C, Vollgraf R (2018) Street2fashion2shop: Enabling visual search in fashion e-commerce using studio images. In: Marsico MD, GS di Baja, ALN Fred (eds) Pattern recognition applications and methods - 7th international conference, ICPRAM 2018, January 16–18, 2018, Revised Selected Papers, Lecture Notes in Computer Science, vol. 11351, pp 3–26. Springer, Madeira, Portugal (2018). https://doi.org/10.1007/978-3-030-05499-1_1

23. Lin S, Hanrahan P (2013) Modeling how people extract color themes from images. In: Proceedings of the SIGCHI conference on human factors in computing systems, CHI '13, p. 3101–3110. Association for Computing Machinery, New York, NY, USA (2013). https://doi.org/10.1145/2470654.2466424

24. Liu Y, Li SZ, Wu W, Huang R (2013) Dynamics of a mean-shift-like algorithm and its applications on clustering. Inf Process Lett 113(1):8–16
25. Lloyd SP (1982) Least squares quantization in pcm. IEEE Trans Inf Theory 28:129–136
26. Lücke, J, Forster D (2019) k-means as a variational em approximation of gaussian mixture models. Pattern Recogn Lett 125, 349–356
27. Manfredi M, Grana C, Calderara S, Cucchiara R (2013) A complete system for garment segmentation and color classification. Mach Vis Appl 25:955–969
28. SizerMe (2020) https://sizer.me/
29. Yazici VO, van de Weijer J, Ramisa A (2018) Color naming for multi-color fashion items. In: Rocha Á, Adeli H, Reis LP, Costanzo S (eds) Trends and advances in information systems and technologies. Springer International Publishing, Cham, pp 64–73
30. Cheng Yizong (1995) Mean shift, mode seeking, and clustering. IEEE Trans Pattern Anal Mach Intell 17(8):790–799
31. Yu L, Zhang L, van de Weijer J, Khan FS, Cheng Y, Párraga CA (2017) Beyond eleven color names for image understanding. Mach Vis Appl 29:361–373
32. ZeeKit (2020) https://zeekit.me/
33. Zhang Q, Xiao C, Sun H, Tang F (2017) Palette-based image recoloring using color decomposition optimization. IEEE Trans Image Process 26:1952–1964

User Aesthetics Identification for Fashion Recommendations

Liwei Liu, Ivo Silva, Pedro Nogueira, Ana Magalhães, and Eder Martins

Abstract One of the challenges in fashion recommendations is how to incorporate the concepts of fashion and style to provide a more tailored personalized experience for fashion lovers. Despite that these concepts are subjective, our fashion experts at Farfetch have defined a few key sets of aesthetics which attempt to capture the essence of users' styles into groups. This categorization will help us to understand the customers' fashion preferences and hence guide our recommendations through the subjectivity. In this paper, we will demonstrate that such concepts can be predicted from users' behaviors and the products they have interacted with. We not only compared a popular machine learning algorithm—Random Forest with a more recent deep learning algorithm—Convolutional Neural Network (CNN), but also looked at 3 different sets of features: text, image, and inferred user statistics, together with their various combinations in building such models. Our results show that it is possible to identify a customer's aesthetic based on this data. Moreover, we found that the use of the textual descriptions of products interacted by the customer led to better classification results.

L. Liu · I. Silva · P. Nogueira (✉) · A. Magalhães · E. Martins
Farfetch, London, UK
e-mail: pedro.nogueira@farfetch.com

L. Liu
e-mail: liwei.liu@farfetch.com

I. Silva
e-mail: ivomiguel.silva@farfetch.com

A. Magalhães
e-mail: ana.magalhaes@farfetch.com

E. Martins
e-mail: eder.martins@farfetch.com

© The Author(s), under exclusive license to Springer Nature Switzerland AG 2021
N. Dokoohaki et al. (eds.), *Recommender Systems in Fashion and Retail*,
Lecture Notes in Electrical Engineering 734,
https://doi.org/10.1007/978-3-030-66103-8_3

Fig. 1 Street aesthetic and feminine aesthetic wear examples

Street

Feminine

1 Introduction

Recommender systems have been an increasingly important part in e-commerce. Some websites are designed to follow a personalized recommendation experience flow [4], such as Netflix (which reported that 75% of the views originate from their recommendations) and Amazon (with 35% of the revenue coming from personalized recommendations). In fact, recommender systems have played such a role in a customers' shopping journey, that today they expect to see recommendations during their interaction with e-commerce websites.

Performing fashion recommendations poses a challenge for those who need to reflect a customer's unique sense of style and preferences in order to enhance a personalized experience for a fashion lover. When a customer visits the website, we should aim to recommend products considering their style preferences and current demand. Recent studies [8] showed that incorporating a user's style into recommendations had mitigated the popular item bias problem in some recommendation domains.

Moreover, in fashion we need to inspire the customer to discover pieces that resonate with their own style and preferences. This has been a mission at Farfetch, an online luxury fashion retail platform that sells products from thousands of boutiques and brands across the world.

Our fashion experts have defined a set of key aesthetic concepts aiming to reflect our current and target customers' style trends. For females, there are six aesthetics: Arty, Classic, Edgy, Feminine, Minimal and Streetwear. While for males, four aesthetics: Edge, Minimal, Smart and Streetwear (examples in Fig. 1). Each of these themes have a product listing page and when customers navigate the page they can start to browse a list of products associated with each of the aesthetics.

This is a great way to capture a customer's style interests. Based on this data, we build models to identify customer's aesthetics and extend the predictions to the rest of our customer base. Once we know which aesthetics a customer belongs to, we not only can identify our customer "neighbors" more accurately, but also can suggest products more tailored to their aesthetic(s). Similarly to [8], we could add this aesthetic as a feature for our recommendation engine.

In this paper, we present how we can identify customers' aesthetics from their online shopping behaviors and the products they have shown interests in. We compare the performance of our models when using 3 completely different sets of features, namely: text, image, and inferred user statistics. We, also, looked at different combinations of those features. On top of this, we explore the performance difference in using a popular classification algorithm—Random Forest (RF), and a more recent deep learning algorithm—Convolutional Neural Network.

We defined this classification problem as a multi-label classification, since customers can show their interests to multiple aesthetics. For example, a customer can associate thyself with both being Classic and Minimal. When a customer is purchasing for others, the aesthetic interests can be very different, they can even be aesthetics from a different gender as well.

Our results show that it is possible to identify a customers' aesthetic based on their navigational patterns. Moreover, we found that the use of the textual descriptions of products interacted by the customer led to better classification results.

The main contributions of our work are twofold: (1) a comprehensive characterization of fashion customers based on their behavior on our platform; (2) a comparison of various models over a rich set of features, capable of classifying a customer into a predefined aesthetic.

The rest of this paper is as follows. Section 2 discusses some related work, while Sect. 3 defines our methodology. Our experimental evaluation is discussed in Sect. 4. Section 5 summarizes the paper and outlines our future work.

2   Related Work

Multi-label classification is the task of assigning a subset of predefined categories to a given item. Classical approaches are based on binary relevance learning (i.e., construct a binary classifier for each category) [3] or a label powerset, by transforming the problem into a multi-class problem with one multi-class classifier trained on all unique label combinations found in the training data [10]. There is more effort focusing on using deep neural networks in recent works. Nam et al. [17] show that a simple NN model trained using cross entropy loss performs, as well as, or even outperforms, state-of-the-art approaches on various textual datasets. Liu et al. [11] present a deep learning approach, based on a Convolutional Neural Network (CNN) model tailored for multi-label classification to tackle the problem of Extreme multi-label text classification (when the number of labels is very high). They show that the proposed CNN approach is scalable to large datasets, and produce competitive to superior results with other state-of-the-art in literature. Here, we compare both classical methods and NN ones in our fashion domain, discussing the pros and cons of each one.

Incorporating user's style into recommendations has been delivering promising results on mitigating popular item bias, for example, Iqbal et al. [8] incorporates user style into a Variational Autoencoder recommendations framework, and found that this addition allowed more diverse recommendations while maintaining relevance in e-commerce context. There are, also, some studies which tackle the problem of assigning a style to a cloth image [5, 7, 12, 20]. For example, Hadi et al. [12] created a crowd sourced dataset to classify clothes according to five different styles. Hsiao et al. [7] propose an unsupervised approach to learn a style-coherent representation for items. The method leverages probabilistic polylingual topic models based on visual attributes to discover a set of latent style factors. Unlike them, in this work, we aim to assign a style to a user, not an image. Other works [9, 13, 23] focused on building representations for items that capture somehow the style of the clothes. We focused on understanding the user and their aesthetic preferences.

3   Methodology

Out of many algorithms that work on multi-label classification, we have selected Random Forest which has been a popular choice and we trusted it to give us a good baseline, and a deep learning model with CNN as in Liu et al. [11], which has shown promising results.

Since aesthetics are gender-specific, and also a customer could be purchasing products for others, especially from another gender, we have decided to model customers based on the product gender they have interacted with regardless of the customers' gender. As a result, a customer can be modelled for both Female and Male aesthetics. Unfortunately without purchase context or intention data, we are not able to divide the modelling into a more refined manner. Those kinds of context data are very hard to obtain in an e-commerce environment without disturbing customers' shopping experience.

Having those two main algorithms in mind and a pool of customers with indication of their aesthetic preferences, the rest is an open question on how to choose the training features for the prediction and what variations to explore to obtain the best offline model possible. In this work, we tried 3 different sets of features:

- Users statistics, the categories and brands they have interacted with
- Image embeddings
- Word embeddings

The rest of this section will explain in detail how we used each feature with each of our two classification algorithms.

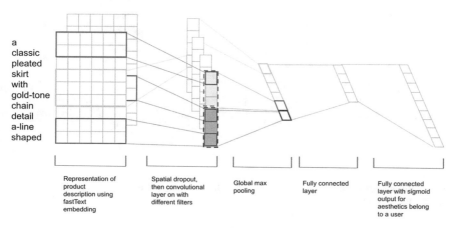

Fig. 2 Architecture used in this paper, graph adapted from [11]

3.1 General Users Statistics Model

In the users' statistics type of features, we have aggregated features per customer and applied the Random Forest algorithm to build a classifier. The features used include: the number of sessions, number of clicks (within six months), number of orders (within two years), number of returned items, the average discount a customer's purchases, the total gross margin values, etc. We also included the categories and the brands a user has interacted within the selected date range. Bearing in mind the curse of dimensionality problem, as the categories and brands data suffer a typical long-tail problem (i.e., the most popular categories cover the majority of the clicks). In this work, we only used the top 100 most popular categories, and the top 100 most popular brands as features.

When using categories or brands as features, their values are normalized weighted actions counts, here we use a category as an example:

$$V(u, cat) = \frac{\sum_{a \in all\,Actions} w_a * Count(a, u)}{\sum_{a \in all\,Actions} w_a} \tag{1}$$

where, u is representing the customer, a is a particular action (i.e. click, order, add to wish list, add to bag, return) and $Count(a, u)$ the count of how many times the customer performed this action. Each of the actions has a weight w_a based on its importance on our platform. Generally, the weight for click action is the smallest, and order action is the largest.

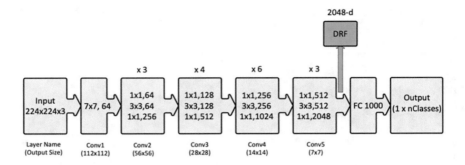

Fig. 3 ResNet50 architecture with residual units, the size of filters and the outputs of each convolutional layer [6, 15]

3.2 Image Embedding Model

CNN have been successful in solving computer vision problems in recent years [6, 21, 22]. There are a few well-known network architectures such as VGG16, VGG19 [21], ResNet50 [6], Inception V3 [22] pre-trained on the ImageNet dataset. In this work, we used the last convolutional layer of ResNet50 for the image feature extraction due to our previous success in other projects in practice. All the product image embeddings are represented as a 1×2048 dimensional vector (Fig. 3).

Here, image embeddings were treated in different ways in order to extract relevant information so that they can be used as features. First, we tried to aggregate all the products image embeddings together per user, i.e., calculating the average, minimum, maximum or quantile of the embeddings of all the products a user has interacted with. This will finally provide a 2048 vector per user, which is then used as a feature.

We also tried to cluster all the products within each category when using image embeddings, on the assumption that the products of a category belonging to a certain aesthetic may share some visual similarity. For example, 3 clusters were formed under the Tops category, and named as cluster_tops_1, cluster_tops_2, and cluster_tops_3, and these 3 names will be features for the Tops category. Hence, the final features will be all the clusters out of all the categories, and the values of these features per user are whether that user has interacted with a product from that cluster or not. In this work, K-means clustering and also variations of the image embedding dimensions using PCA are also experimented. Both sets of image features are used as input for Random Forest algorithm.

3.3 Word Embedding Model

Product descriptions are used as another alternative to generate features on the assumption that there would be some indication of the style of products from the way they are described. We first tried to use term frequency—inverse document frequency (TF-IDF) to prepare the values of each token. The tokens are generated through a series of NLP processes such as converting words to lower case, removing punctuation, stemming and finally transforming to tokens.

In the next iteration, FastText [1] was used to generate the embeddings for each word. Since most brands in our text are non frequent occurrences, FastText is a better choice in comparison with word2vec [16] or GloVe [19]. FastText represents words as the sum of a bag of characters of n-gram.

FastText vectors are then used as input to Random Forest and we also trained a CNN model using them. The architecture for training this neural network can be observed in Fig. 2.

The model consumes the embedding of the words/tokens from each product description out of all the interacted products per user using a pre-trained FastText model, which is followed by a spatial dropout and a convolutional layer with different filters. Then a global max pooling layer, which was flattened out and passed to a fully connected layer. Finally, we reach the output layer corresponding to the number of aesthetics. The multi-label loss function following the Eq. 1 [11]:

$$\min_{\Theta} -\frac{1}{n} \sum_{i=1}^{n} \sum_{j=1}^{L} y_{ij} \log\left(\hat{p}_{ij}\right) = -\frac{1}{n} \sum_{i=1}^{n} \sum_{j \in y_i^+} \frac{1}{|y_i^+|} \log\left(\hat{p}_{ij}\right) \tag{2}$$

where Θ represents model parameters, y_i^+ represents the set of relevant labels of instance i and $\left(\hat{p}_{ij}\right)$ is the model prediction for user i on label j.

4 Experiments and Results

4.1 Dataset and Evaluation

Each of the aesthetic concepts has its own URL link which will lead to a list of products associated with that aesthetic on Farfetch website. We defined that a customer is interested in some aesthetic when they navigated to an aesthetic listing page and then proceeded to click in at least one of the listed products, considering the last six months period. For simplicity, only information about Female Aesthetics are shown here. As you can see in Fig. 4, the total number of users that interacted with each of the aesthetic concepts is quite balanced, with Streetwear having the highest number of users (more than 7000) and Artistic the lowest numbers of users (a few more than 5000).

Fig. 4 Total number of users that interacted with each female aesthetic concept

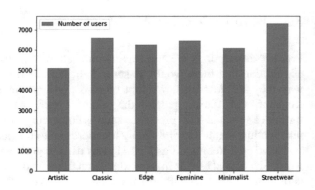

Fig. 5 Total number of users by number of aesthetic concepts they interacted with

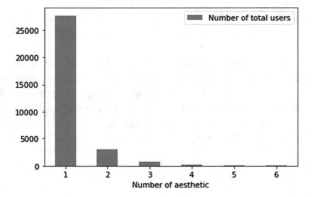

There are 31,900 users that clicked in aesthetic listing pages from female gender and interacted with the products presented in those pages. Typically, 50% of those users have done more than 230 actions that can be just clicking in the product, adding it to the wishlist/bag or purchasing it.

Nonetheless, most of those actions seem to be related to products belonging to the same aesthetic concept as, on average, each user has interacted with 1.18 aesthetic concepts, as shown in Fig. 5. In fact, more than 85% of the users only interacted with one aesthetic and respective products.

Moreover, as shown in Fig. 6, it is clear that the aesthetic concepts do not have any correlation between them, i.e. the labels are independent of each other. This proves to be very important when it comes to the modelling task because it supports the idea of using multiple binary classification models (one for each label), rather than training a single multi-label model.

We randomly splitted our data in 75% for training and 25% for testing, in a 5-fold cross-validation setting. Precision, Recall and F1 were used as the evaluation metrics. In the case of binary models, we used micro aggregation on those three metrics.

Fig. 6 Correlation between aesthetic concepts (labels)

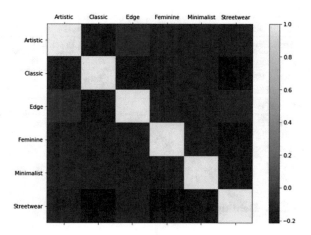

4.2   Discussion

Experiments were carried out testing on a few main variations, aiming to help us to understand what could be the best model for productization. Only the results from modelling Female Aesthetics are shown here, for the interest of clarity. The Male Aesthetics results achieved similar outcomes when it comes to modelling purposes.

4.2.1   Multi Label Configuration

We conducted preliminary experiments to determine what is the best technique to approach our multi label problem. We tested with both classical approaches (binary relevance and label powerset) and, also, with the default *scikit-learn* implementation of random forests (that supports multi label classification[1]).

Overall, models trained using binary relevance strategy performed slightly better than models that employed the power set technique. To our surprise, we noted that in some cases, the *scikit-learn* multi-label model had very odd performance. For example, when using TF-IDF features, the recall is almost 1, i.e. almost all the predictions are 1 for all classes. This indicates that just averaging impurity reduction across all the outputs could not be enough to build a reliable multi label model for our scenario. So, we decided to use binary relevance for training all the Random Forest models.

---

[1]It builds a single generalized model capable of processing output correlations. To build a tree, it uses a multi-output splitting criteria computing average impurity reduction across all the outputs. To the best of our knowledge, this could be viewed as a kind of greedy label powerset technique.

Table 1 Evaluation results when applying different treatments for data imbalance with general stats in binary Random Forest classification models

Treatment	F1	Precision	Recall
No treatment	0.043	0.771	0.022
Class_Weight Balanced	0.442	0.360	0.572
SMOTE	0.400	0.345	0.476
SMOTE combined with Class_Weight Balanced	0.402	0.347	0.478

4.2.2 Dealing with Data Imbalance

When creating a classifier for each class, our dataset may become imbalanced, as all other classes will represent negative samples. As shown in Fig. 5, the percentage of users having more than one aesthetic is small, less than 15% of the whole population. Therefore, 85% of the samples are considered the negative class. This could cause imbalance problems models, so we tested with some strategies to deal with this problem.

We first tried to use the "class_weight" parameter in Random Forest implementation to automatically adjust the class weights inversely proportional to class frequencies in the input data [18]. We also tried to use SMOTE [2] which aims to create synthetic data to help to reduce the data imbalance problem. Our results, summarized in Table 1, confirm that there is a significant improvement when we balanced the dataset, either using class weight or SMOTE. In fact, the first approach seems to produce better results. We did try different options under the SMOTE domain, all of which produced very close results. This set of experiments were all carried out using general stats features in Random Forest binary classifiers, although very similar results could be obtained in each of the feature sets.

4.2.3 Impact of the Choice of Training Features

This set of experiments looks into how the same model performs under different feature sets. We trained one model for each feature set combination for female aesthetics using Random Forest. We also trained an additional Random Forest model over a random generated feature set. This could be seen as a lower bound for our metrics (in fact, as expected, all the models outperformed the random one). Table 2 shows the approximate size of each feature set.

Table 3 shows our results. In general, word embedding (TF-IDF) performs better than other models using a single feature set, without any dimensionality reduction. From the reasoning of features choices, it makes sense that features generated from product descriptions perform well, since some words could be a strong indication of

Table 2 Approximate size of each feature set

Feature set	Approximate size
General stats	300
Image embedding (clusters)	2048
Word embedding (TF-IDF)	455K
Word embeddings (FastText)	300

Table 3 Evaluation results when using different sets of training features with binary Random Forest classification models

Features	F1	Precision	Recall
Word embedding (TF-IDF)	0.525	0.586	0.476
Word embedding (TF-IDF) + Image embedding	0.524	0.585	0.474
General Stats + Word embedding (TF-IDF) + Image embedding	0.507	0.513	0.500
General Stats + Word embedding (TF-IDF)	0.503	0.521	0.486
General Stats	0.442	0.360	0.572
General Stats + Image embedding	0.420	0.382	0.467
Word embeddings (FastText)	0.418	0.335	0.554
Image embedding (clusters)	0.348	0.295	0.424
Random	0.257	0.197	0.370

aesthetics. Interestingly, when using FastText the results decrease, this could indicate that there are some words with a special meaning in the fashion domain that have not been captured by an embedding model trained over a more general text dataset. An in depth analysis of this fact is out of scope of our objective and is defined as a future work.

The general stats feature model results seem promising, particularly on recall, indicating that users in a particular customers' aesthetics share some navigation patterns.

On the other hand, models based on image features are performing the worst. The main reason might be because in the same aesthetic concept those features can be too generic since the products can look very different, i.e. with different patterns, different shapes, etc.

Finally, some variations we tried seemed to show no sensitivity in the outcome, as we can see in the image embedding and general stats combination (Table 3). This could indicate that those features do not have complementary information that could be exploited by the Random Forest algorithm.

Table 4 Evaluation results when using Random Forest comparing to using a CNN deep learning model with different time length in multi-label classification

Data time range	Features	Algorithm	F1	Precision	Recall
6 months	Word embedding (TF-IDF)	RF	0.525	0.586	0.476
3 months	Word embedding (TF-IDF)	RF	0.505	0.555	0.463
6 months	Word embedding— FastText	CNN	0.404	0.680	0.288
3 months	Word embedding— FastText	CNN	0.307	0.687	0.199

4.2.4   CNN Results

Deep learning models seem to be another promising way of modelling, especially with the recent multi-label classification development [11]. With that in mind, we compared Random Forest with CNN (Sect. 3.2) in a multi-label classification setting. To do so, we chose the best Random Forest model trained over a single feature set— Word embedding (TF-IDF)—and tested it against the CNN model trained on the same feature set.

As you can see on Table 4, the deep learning models tend to need more data, we tested using the last 3 and 6 months of products that a user has interacted with. This increase in data had little impact on Random Forest, but helped CNN to significantly improve recall. In general, the results seem promising. CNN models performed better on precision at the cost of recall leading to a worst F1 when compared with Random Forest. It is possible that with experiments using a bigger dataset the CNN model can be further improved. Also, our CNN model is not fully optimized, in particular, we did not conduct a deep study on the impact of the class imbalance on the CNN model which may affect its performance. In future iterations, we plan on exploring different techniques to tackle this problem.

4.2.5   Different Loss Functions

We have hypothesised over using binary cross entropy or categorical cross entropy as the loss function in a multi-label classification setting. The authors in [11] used binary cross entropy, on the other hand [14] mentioned that categorical cross entropy seems to perform better. We tried both in our CNN model and, in our case, the binary cross entropy loss function had a better F1 score (Table 5).

Table 5 Evaluation results when using different loss functions for CNN deep learning model trained using word embedding—FastText

Loss function	F1	Precision	Recall
Binary cross-entropy	0.404	0.680	0.288
Categorical cross-entropy	0.284	0.702	0.178

Table 6 Evaluation results breakdown by Female Aesthetic for Random Forest with word embeddings (TF-IDF)

Aesthetic	Frequency	F1	Precision	Recall
Streetwear	0.230	0.635	0.739	0.557
Classic	0.207	0.589	0.671	0.524
Feminine	0.203	0.483	0.554	0.428
Edge	0.196	0.484	0.551	0.431
Minimalist	0.192	0.499	0.583	0.437
Artistic	0.158	0.442	0.525	0.383

4.2.6 Aesthetic Result Breakdown

In order to evaluate how well the best model—Random Forest with word embeddings (TF-IDF)—perform in each Aesthetic label, we present the F1, Precision, and Recall on Table 6. We can see that there is a positive correlation between class frequency and the evaluation metrics which might be an indicator that more popular aesthetics are easier to classify.

As you can see from the example of Fig. 7, a user that was interested in the products presented was correctly labeled as belonging to the Feminine Aesthetic. Looking at the products, we can say that they look like very feminine products, however, the images are very diverse between each other. This observation, might sustain the fact that image embeddings do not provide useful information to the models when compared with other features.

On the other hand, if we look at the product descriptions, we can see that there are some words in common between products like "midi" and "dress" that were crucial for the model to classify the Aesthetic correctly.

5 Conclusion and Future Work

We have explored identifying customers' style preference through aesthetic concepts in various ways. We demonstrated that those aesthetics could be inferred from the customer's online shopping behaviors and the products they have shown interests in. In the end, using Random Forest with binary relevance to tackle this multi-

cowl neck tank mini dress (W / 0)   puff sleeve midi dress (W / 0)   Livarno satin midi skirt (W / 0)   June v-neck camisole (W / 3)   Kim high neck top (W / 19)
Maisie Wilen   **CECILIE BAHNSEN**   **GAUGE81**   **BERNADETTE**   **ROTATE**

Fig. 7 List of products that a user, correctly labeled as Feminine Aesthetic, interacted with

label problem seems like the best option for the dataset we currently have available. Also, using text features generated from product description seems to have a better performance when compared with other feature sets (image embedding, user general stats).

As a future work, we will carry out a customer survey aiming to collect more data on our customers perception about their aesthetics. With this type of data, we can evaluate our models with customer survey data to see whether the results will align with what we have found in this paper. Moreover, we will further explore our CNN model, in particular, we want to study the class imbalance problem and the reason why FastText embedding performs worse than the TF-IDF feature set.

References

1. Bojanowski P, Grave E, Joulin A, Mikolov T (2016) Enriching word vectors with subword information. arXiv preprint arXiv:1607.04606 (2016)
2. Chawla NV, Bowyer KW, Hall LO, Kegelmeyer WP (2020) Smote: Synthetic minority over-sampling technique. arXiv preprint arXiv:1106.1813 (2020)
3. Dembczynski K, Cheng W, Hüllermeier E (2010) Bayes optimal multilabel classification via probabilistic classifier chains. In: Fürnkranz J, Joachims T (eds) ICML, pp 279–286. Omnipress (2010). http://dblp.uni-trier.de/db/conf/icml/icml2010.html#DembczynskiCH10
4. Faggella D (2017) The roi of recommendation engines for marketing. https://martechtoday.com/roi-recommendation-engines-marketing-205787 (2017). Accessed 29 July 2020
5. Gonçalves D, Liu L, Magalhães A (2019) How big can style be? addressing high dimensionality for recommending with style. CoRR abs/1908.10642. http://dblp.uni-trier.de/db/journals/corr/corr1908.html#abs-1908-10642
6. He K, Zhang X, Ren S, Sun J (2015) Deep residual learning for image recognition. arXiv preprint arXiv:1512.03385
7. Hsiao WL, Grauman K (2017) Learning the latent "look": Unsupervised discovery of a style-coherent embedding from fashion images. In: ICCV, pp 4213–4222. IEEE Computer Society. http://dblp.uni-trier.de/db/conf/iccv/iccv2017.html#HsiaoG17
8. Iqbal M, Aryafar K, Anderton T (2019) Style conditioned recommendations. CoRR **abs/1907.12388**. http://dblp.uni-trier.de/db/journals/corr/corr1907.html#abs-1907-12388
9. Lee H, Seol J, Lee S (2019) Style2vec: Representation learning for fashion items from style sets. CoRR **abs/1708.04014**. http://arxiv.org/abs/1708.04014
10. Liu F, Zhang X, Ye Y, Zhao Y, Li Y (2015) Mlrf: Multi-label classification through random forest with label-set partition. In: Huang DS, Han K (eds) Advanced intelligent computing theories and applications. Springer International Publishing, Cham, pp 407–418

11. Liu J, Chang WC, Wu Y, Yang Y (2017) Deep learning for extreme multi-label text classification. pp 115–124. https://doi.org/10.1145/3077136.3080834
12. M. Hadi Kiapour Kota Yamaguchi ACBTLB (2014) Hipster wars: Discovering elements of fashion styles. In: European conference on computer vision (2014)
13. Magalhães AR (2019) The trinity of luxury fashion recommendations: data, experts and experimentation. In: Bogers T, Said A, Brusilovsky P, Tikk D (eds) RecSys, p 522. ACM. http://dblp.uni-trier.de/db/conf/recsys/recsys2019.html#Magalhaes19
14. Mahajan D, Girshick R, Ramanathan V, He K, Paluri M, Li Y, Bharambe A, van der Maaten L (2018) Exploring the limits of weakly supervised pretraining. arXiv preprint arXiv:1805.00932
15. Mahmood A, Ospina AG, Bennamoun M, An S, Sohel F, Boussaid F, Hovey R, Fisher RB, Kendrick G (2020) Automatic hierarchical classification of kelps using deep residual features. arXiv preprint arXiv:1906.10881
16. Mikolov T, Chen K, Corrado G, Dean J (2013) Efficient estimation of word representations in vector space, 1–12. http://arxiv.org/abs/1301.3781
17. Nam J, Kim J, Loza Mencía E, Gurevych I, Fürnkranz J (2014) Large-scale multi-label text classification – revisiting neural networks. In: Calders T, Esposito F, Hüllermeier E, Meo R (eds) Machine learning and knowledge discovery in databases. Springer, Berlin Heidelberg, Berlin, Heidelberg, pp 437–452
18. Pedregosa F, Varoquaux G, Gramfort A, Michel V, Thirion B, Grisel O, Blondel M, Prettenhofer P, Weiss R, Dubourg V, Vanderplas J, Passos A, Cournapeau D, Brucher M, Perrot M, Duchesnay E (2011) Scikit-learn: Machine learning in Python. Journal of machine learning research 12:2825–2830
19. Pennington J, Socher R, Manning CD (2014) Glove: global vectors for word representation. In: Empirical methods in natural language processing (EMNLP), pp 1532–1543. http://www.aclweb.org/anthology/D14-1162
20. Simo-Serra E, Ishikawa H (2016) Fashion style in 128 floats: Joint ranking and classification using weak data for feature extraction. In: CVPR, pp 298–307. IEEE computer society (2016). http://dblp.uni-trier.de/db/conf/cvpr/cvpr2016.html#Simo-SerraI16
21. Simonyan K, Zisserman A (2014) Very deep convolutional networks for large-scale image recognition. arXiv preprint arXiv:1409.1556
22. Szegedy C, Vanhoucke V, Ioffe S, Shlens J, Wojna Z (2015) Rethinking the inception architecture for computer vision. arXiv preprint arXiv:1512.00567
23. Veit A, Kovacs B, Bell S, McAuley JJ, Bala K, Belongie SJ (2015) Learning visual clothing style with heterogeneous dyadic co-occurrences. In: ICCV, pp 4642–4650. IEEE computer society. http://dblp.uni-trier.de/db/conf/iccv/iccv2015.html#VeitKBMBB15

Sizing and Fit in Online Fashion

Towards User-in-the-Loop Online Fashion Size Recommendation with Low Cognitive Load

Leonidas Lefakis, Evgenii Koriagin, Julia Lasserre, and Reza Shirvany

Abstract One of the major challenges facing e-commerce fashion platform is that of recommending to customers the right size and fit for fashion apparel. In this work we study this topic in depth and demonstrate its various complexities focusing in particular on the challenging cold-start problem that arises when no order history is available for a specific customer. We demonstrate the multifaceted value of data obtained by involving the customer in the loop and show how it allows for an effective cold-start recommender system. We highlight our findings via detailed experiments performed on hundreds of thousands of customers and items in real world e-commerce scenarios. In addition, results and discussions are provided investigating the trade-off between the recommender's effectiveness and the customer's experience with the goal of introducing accurate solutions with low user cognitive load.

1 Introduction

Finding fashion apparel online with the right size and fit is challenging for many customers. It is actually one of the major factors impacting not only customers purchasing decisions, but also customers satisfaction with e-commerce fashion platforms. The underlying difficulties mean that either customers remain reluctant to engage in the purchasing process, in particular with regards to new articles and brands they are not familiar with, or they purchase articles in multiple neighboring sizes to try them out and return the ones that are not fitting. Compounding the issue, customer prefer-

L. Lefakis (✉) · E. Koriagin · J. Lasserre · R. Shirvany (✉)
Zalando SE, Berlin, Germany
e-mail: leonidas.lefakis@zalando.de

R. Shirvany
e-mail: reza.shirvany@zalando.de

E. Koriagin
e-mail: evgenii.koriagin@zalando.de

J. Lasserre
e-mail: julia.lasserre@zalando.de

© The Author(s), under exclusive license to Springer Nature Switzerland AG 2021
N. Dokoohaki et al. (eds.), *Recommender Systems in Fashion and Retail*,
Lecture Notes in Electrical Engineering 734,
https://doi.org/10.1007/978-3-030-66103-8_4

ences towards perceived article size and fit for their body remain highly personal and subjective which in turn influences the definition of the right size for each customer.

In order to achieve these goals, major fashion platforms are experimenting with providing size and fit advice to steer customers' behavior using a variety of approaches. One of the simplest methods used is that of size tables and aggregated article measurements [1] provided per brand and article category. This approach requires customers to find what fits them best themselves and relies on various body measurements, such as "bust", "waist", "hip", also typically measured by the customers themselves. However, these charts rarely help a customer select the best size [2]. Another emerging approach is that of directly addressing customers, for example via questionnaires or dialogue windows [3] to provide a size advice based on customers' explicit data. Such approaches have been recently adopted by major e-commerce platforms [4–6], and require customers to explicitly provide personal information, such as age, weight, height, tummy shape, hips form, body type, favorite brand, fit preferences (such as slim vs. normal), customers usual sizes, and so on, used to provide size advice for the customers. In a similar vein of asking customers for explicit body data, computer vision and 3D approaches [7–10] have also shown promising results in providing virtual-fit advice. Such solutions require customers to submit, typically high definition images or videos of their bodies, in predefined poses and tight fitting clothes. Such strict requirements are necessary to allow the currently available technology to infer relatively accurate body measurements, size and shape. In contrast to these approaches are the recent work that do not require any explicit data from customers to provide size advice but rather exploit customers rich order history in order to infer the appropriate size advice [11–17] on future orders.

Each of these various approaches typically rely on disparate assumptions making each appropriate for different customer segments and experiences. In particular it is obvious that the data used by each of these approaches is very different in nature and each require a different level and type of engagement from the side of the customer. Certainly a comprehensive comparison of all these approaches would be a high-value to the community- this remains out of the scope of this paper and constitutes a great future research direction. Here we aim to investigate the size recommendation problem in the so called cold-start scenario in which there is little to no order history for each customer, and thus, customers are part of the solution by providing explicit information through some sort of questionnaire. We also aim to create a solution that comes with a low cognitive load for the customers in a way that we burden the customers as little as possible while providing quality size advice.

The contributions of this work are 4-fold: 1. We analyze the sizing characteristics of apparel from hundreds of brands available in fashion e-commerce context and formulate major challenges faced in building large-scale reliable size recommender systems with or without order history; 2. We demonstrate the multifaceted value of customer metadata in effectively tackling the cold-start problem by leveraging said data to build an effective cold-start recommender comparable with those state-of-the-art solutions that have privilege access to customers order data; 3. We leverage the trade-off between the size recommender effectiveness and the required customer data and propose, for the first time to our knowledge, an accurate large-scale size recommender with low customer cognitive load which has been rolled out in six European countries covering various size system conventions. 4. With experimental evidence we furthermore demonstrate how current state-of-the-art size recommendations benefit from our findings from the cold-start problem even in a hot-start setting.

2 Complexity of the Size and Fit Problem at Scale

In order to highlight the scale of the size recommendation problem, we analyzed fashion articles and orders in the category of Female Upper Garments which encompasses a large variety of different fashion apparel, from dresses to denim jackets, and is strongly representative of the complexity of fashion in general and of the many obstacles that arise in personalized size recommendations in particular. In Fig. 1 we present a bar plot of the number of distinct apparel sizes, in the female upper garment category, from around 2000 brands available on a large-scale e-commerce fashion platform during the 2015–2019 time period. In this plot we see that when we aggregate the list of all possible sizes for all brands (composed of all the different size systems such as numeric 38–39-40, ... standard S,M,L, ..., fractions 41 1/3, 42.5, ... confection sizes 36–38, 40–42, ... country conventions EU, FR, IT, UK ...), we reach the upper bound of 17k sizes. We also see that the scale of the size recommendation and size selection problem has grown continuously and rapidly with the number of distinct sizes more than doubling within this category over last 4 years. Diving deeper, Fig. 2 represents the bar plot of the number of distinct sizes per brand for 80 popular brands. Here each vertical bar represents the number of sizes offered by one distinct brand. Brands create these distinct sizes due to multiple underlying (and often undisclosed) business and product optimization rationals [2, 18]. We can see that, already in this category, some brands offer upward of 30 different distinct sizes, leaving customers to face a challenging decision with regards to which exact size

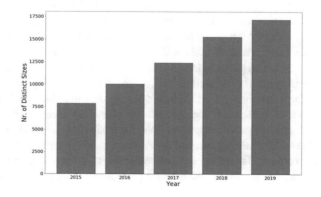

Fig. 1 Bar plot of the number of distinct apparel sizes, in women upper garment category, from hundreds of brands with different size systems and country sizing conventions during 2015 to 2019 time period. The scale of the size recommendation and selection problem continues to grow rapidly in recent years

Fig. 2 Bar plot of the number of distinct sizes per brand, in women upper garment category, for 80 major brands in e-commerce fashion. Each vertical bar represents one distinct brand

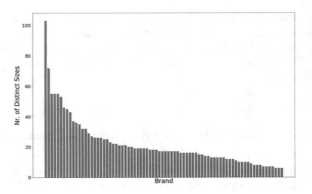

to select when shopping within those brands, and most notably what this means for them when shopping those brands with much fewer sizes to select from.

As fashion e-commerce is increasingly growing, assisting customers in buying the right size presents a huge opportunity for research in intelligent size and fit recommendation systems which can directly contribute to increasing customer satisfaction, reducing environmental footprint, and helping business profitability.

3   Prior Work

Although customer-centric product recommendation is a well-studied field (see [19–23]), size recommendation is still in its infancy with only a few approaches addressing parts of this problem during the past few years [7–17, 23–26]. A large family of these approaches depend on historical data of customer orders and returns either using statistical [11–13] or deep-learning [16, 17, 27] methods, often concentrating on finding suitable embeddings to represent customers and orders [15, 24, 25]. Such approaches have the advantage of not asking customers for explicit data, thus involve a low cognitive effort from the customer. However, such solutions invariably suffer from the cold-start problem which affect thousands of existing and new customers visiting the shopping platforms everyday for which no prior order history is available in hundreds of brands and tens of fashion categories. The cold start problem has been widely studied in the context of user item recommendations [26]. Prior work has typically focused either on the article side [17], or on entering a dialogue with the customers [3–9]. The former approaches, i.e. exploiting attributes of articles, allow to alleviate the cold-start problem by using content-based filtering [23] and come with a low cognitive load by design. However, in the context of personalized size recommendations, article data does not bring sufficient information to tackle the problem. The latter approaches come with the advantage of allowing customers to become an direct and integral part of the recommendation system but on the other side either require customers to share a considerable amount of sensitive personal data (such as age, weight, height, tummy shape, hips form, body type, favorite brand, fit preferences, etc.) [4–6] or require multiple high resolution images, videos, and in some cases 3D scans, of customers bodies in tight clothing and canonical poses [7–9]. Such imagery or 3D data is often used to create virtual-fitting solutions using recent 3D human body estimation and reconstruction approaches [28, 29]. As such, these approaches come with a high level of cognitive load for customers by demanding strong engagement and willingness from them to share images and scans of their bodies with fashion platforms.

A comprehensive comparison of these diverse approaches would be of high value to the community, but this remains out of the scope of this paper and constitutes a great future research direction. Here, in particular we focus on the the size recommendation problem in the so called cold-start scenario in which there is little to no order history for each customer. The cold-start problem in personalized size recommendation is a new emerging field and the underlying importance and necessity of requiring a diverse set of personal and body data for providing these recommendations requires deeper discussion and investigation in the fashion recommendation systems. Here we aim to investigate the size recommendation problem in this cold-start scenario, by involving customers in the loop through some sort of questionnaire and to create a solution that comes with a low cognitive load for the customers in a way that we burden the customers as little as possible while providing quality size advice.

4 Size Recommendation Without Order History

We model the problem of cold-start size recommendation as a categorical classification task, where each size is a possible class. The idea is to directly involve customers in the process by leveraging those for which we have both customer data and purchase data to learn a mapping from customer data to ordered sizes, thus allowing us to predict appropriate sizes for any new customer.

Customer Data: We use a comprehensive set of customer data gleaned from questionnaires presented to customers as part of a specialized online fashion styling service wherein customers are paired with professional stylists who then curate personalized outfits for them. These questionnaires cover a wide variety of fashion related areas, and for this study we extract from the customers' answers the subset of information related to size and fit and use it to build a feature representation for each customer. This subset consists of 20 attributes for each customer falling into three categories as can be seen in Table 1. A total of 450k questionnaires are available, each from a distinct customer which self-identified as female. The questionnaire data is projected onto an input space by calculating a vector representation for each customer. Of the 20 size-related questions on the questionnaire, 7 result in categorical variables and are one-hot encoded while the remaining 13 result in numerical variables and are normalized by mean and standard deviation.

Order data: The order data used in this work is composed of roughly 7.4 million orders placed on an e-fashion platform in the female upper body garment category.

Table 1 Features in the Questionnaire Data

Type	Features
Overall Upper body	Weight, height, age, gender top size, shirt collar size, shirt fit, prop. belly, top fit, prop. shoulder-waist, bust number, bust cup size, prop. shoulderhip, blazer size
Lower body	pants size, jeans length, jeans width, prop. waist, pants waist-height , shoe size

This dataset is anonymized and is not public due to various customer privacy challenges and proprietary reasons. We split these orders into training and test sets based on order timestamps. Of these 7.4 million orders, the oldest 5.6 million comprise the training set while the most recent 1.8 million form the test set. All ordered articles are associated with a numerical size in [34, 36, 38, 40, 42, 44, 46]. Both in training and testing, the target variable is the size bought by the customer. There are cases where different target values correspond to the same input vector as customers will at times buy different sizes. Nonetheless allowing the classifier to handle such ambiguity was found to be the best strategy, as opposed for example to using the median or mean size in training. Using the output of the Hot-Start recommender [13] (presented below) as a target value also proved sub-optimal.

Classifier: We experimented with a variety of multi-class classifiers and found Gradient Boosted Trees to perform best in practice. This choice comes with the added benefit of having an easily interpretable classifier, which as we shall show in the following can prove very useful. The hyper-parameters were tuned using a grid-search and cross-validation (splitting the Training Data into Train and Validation sets). In particular we found that performance saturated at 500 trees, and did not observe any over-fitting effect when growing the ensemble beyond this point. Each tree in the ensemble has a depth of 3, while the trees themselves are built, sequentially, using a learning rate of 0.01 and a sub-sampling rate of 0.5. We note that performance on the validation set was largely robust to the latter two hyper-parameters.

Hot-Start recommender baseline: The size recommender system introduced in [13] has shown to be robust and effective in a large-scale fashion e-commerce context; we thus employ it as a Hot-Start recommender baseline (built on order history data). It follows a hierarchical Bayesian approach that models jointly the probability of a size and return status (kept, too small, too big) given a customer and an article. This approach enjoys the advantages of Bayesian modeling, and in contrast with [11, 12] which have to predict the fit (good fit, too big, too small) for all possible sizes one by one, it can directly predict the probability of any size given a customer-article pair, conditionally on those articles being kept (good fit). It naturally fails in the case of new customers or customers with scarce order history. As one might expect however, the order data falls into the long tail problem, and as such, this shortcoming of current Hot-Start recommenders is quite pronounced in the

Fig. 3 Plot of the number of customers versus the number of prior orders. The long tail nature of the problem is evident where the vast majority of customers are cold-start customers with little to no order history

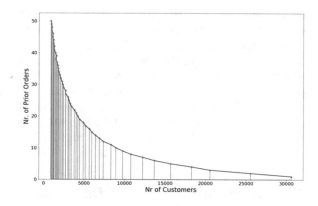

Fig. 4 Brand Size Offsets for the 80 most popular brands on the e-fashion platform. The standard deviation is plotted against the mean

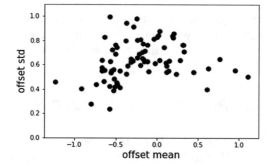

context of online fashion where a large percentage of customers are in the cold-start category with none-to-scarce order history as shown in Fig. 3.

Brand Size Offsets: Most brands suffer a high variance between their nominal and actual sizes caused by multiple design and business related factors, and as such the knowledge of customers' "usual" size alone is often insufficient information for providing both intra- and inter- brand size recommendations. We make use of customers return trends to gain a better understanding of various brands behaviour with respect to size and fit. Customers often tend to order, for a certain brand, their "usual" size and one size up or down, either within the same order or in a later order if they returned the first-ordered size. Therefore, using data regarding kept and returned articles, one can readily define an article offset as the difference in the sizes from the (reordered) kept articles and the ordered (but not kept) articles. An offset for a brand is then defined as the weighted average of all the article offsets in that brand. We weigh the contributions of each article by the number of sales so that the highest selling articles contribute the most to the final offset of the brand, and calculate a weighted mean (μ) and a weighted standard deviation (σ) to fit a Gaussian. We show the range of brand offsets in Fig. 4, where μ and σ are calculated for 80 popular brands using 10k distinct women upper-garment purchases per brand. We highlight the importance of exploiting brand offsets in Sect. 5 where we present our results on cold-start recommendation.

Recommender predictions: The trained classifier is combined with the brand offsets presented above resulting in our Cold-Start recommender. To obtain the final prediction of the Cold-Start recommender, the predicted class c of the trained ensemble is combined with the brand offsets by adding the brand offset mean μ_{brand} to the ensemble prediction and the final size recommendation is obtained by rounding $c + \mu_{brand}$ to the closest size. We note that we use brand offsets for the baseline Hot-Start recommender too, as they were found to be advantageous.

5   Experimental Results and Discussion

In this section, we evaluate the recommender systems based on the accuracy metric, defined as the percentage of times the recommender correctly predicts the size bought by the customer on the orders in the test set.

5.1   Hot-Start and Cold-Start Performances

In Fig. 5 we present a comparative study of the baseline Hot-Start recommender and the proposed Cold-Start recommender. We plot the accuracy of the models against the number of prior orders of each customer in the training data (as we employ temporal split of the data into training and test set, we can also speak of training and test time). As can be seen, for low number of orders regime (<10) the Cold-Start recommender (in blue) clearly outperforms the Hot-Start one (in yellow), even in cases when a customer has a substantial amount of prior orders (>10, <20). We re-iterate that the Cold-Start recommender does not use any knowledge of prior orders. The Hot-Start recommender only starts outperforming the Cold-Start recommender after about 20 prior orders. This is due to the hard limitations of most current Hot-Start recommender approaches where they need a minimum set of orders per each sub-level category to perform, as has been duly noted in [13, 15], effectively restricting Hot-Start solutions to loyal customers with rich order history. Overall the performance is 58% accuracy for the Cold-Start recommender and 54% for the Hot-Start recommender (see Cold-Start (All Data) and Hot-Start (Baseline [13]) in Table 3).

In Fig. 6 we present the confusion matrix of the Cold-Start recommender. We note that, even when wrong, the predictions are seldom off by more than a size. Furthermore the classifier struggles most with the more popular sizes (e.g.. 40, 42). Figure 7 shows the equivalent confusion matrix for the Hot-Start recommender. The same observations as for the Cold-Start recommender can be made, highlighting the inherent ambiguity and complexity of the right recommendation for popular sizes.

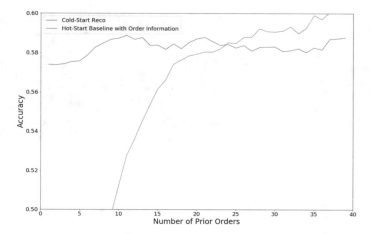

Fig. 5 Comparison of the Cold-Start recommender system presented here to the Hot-Start system presented in [13]. On the x-axis are the number of prior orders (specifically the number of prior orders of each customer in the training set). On the y-axis is the accuracy defined as the percentage of times the recommender correctly predicts the size bought by the customer on the orders in the test set

Fig. 6 Cold-Start recommender's confusion matrix, in black. The subplot below shows the distribution of sizes in sales, and the one on the right hand side shows the accuracy of the Cold-Start recommender per size

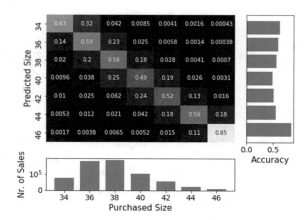

5.2 Impact of Brand Size Offsets

As noted, the proposed Cold-Start recommender makes great use of the brand offsets. To highlight the contribution of these offsets in the performance of the recommender, we present in Fig. 8 the accuracy of a Cold-Start recommender based only on the ensemble method which does not exploit the brand offsets (Cold-Start Reco Without Brand Offsets), and the accuracy of the proposed Cold-Start recommender which adds the brand offset's mean to the ensemble output. We plot these accuracies relative to a lower threshold on the standard deviation, whereby for a given threshold θ we

Fig. 7 Hot-Start
recommender's confusion
matrix, in black. The subplot
below shows the distribution
of sizes in sales, and the one
on the right hand side shows
the accuracy of the Hot-Start
recommender per size

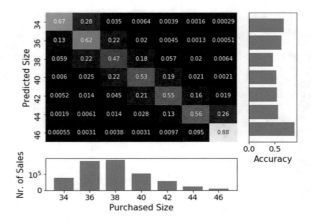

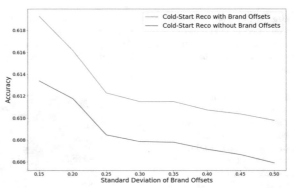

Fig. 8 Accuracy of the
Cold-Start recommender
depending on whether brand
offsets are exploited to refine
size recommendations

only include those brands which have a standard deviation above θ to calculate the
accuracy. As can be seen there is a clear advantage to exploiting brand offsets when
making recommendations.

In Fig. 9 we present the percentage of brands for which we observe improved
accuracy when adding the brand offset means relative to the standard deviation of
the offsets. Again we plot against a lower threshold on the brand standard deviations.
As can be seen in the case of brands with an offset standard deviation of at least
0.15 adding the brand offset means results in improved accuracy in approximately
66% of cases. As expected, the positive effect of using brand offsets diminishes as
the standard deviation rises. As the standard deviation reaches 0.5 the brand offsets
means lead to improved performance only in approximately 33% of cases. To address
this limitation, a future work direction is to directly use the article offsets for brands
suffering from such high standard deviations.

Fig. 9 Percentage of brands
for which adding the brand
offset increases accuracy
plotted against a lower
threshold on the std

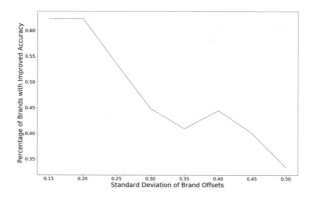

Fig. 10 Comparison of the
Cold- and Hot-Start
recommender systems
relative to the percentage of
customers covered (customer
coverage)

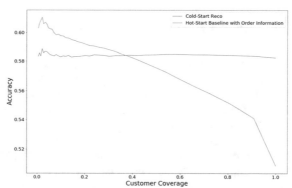

5.3 Customer Coverage

In order to get a better understanding of the percentage of customers covered by
various order segments[1] we plot in Fig. 10 the accuracy of the recommender systems
relative to these percentages. By taking all customers we achieve a customer coverage
of 100% although in this case the Hot-Start Reco performs quite poorly as can be seen
in the plots. On the other hand taking only those customers who have a large number
of prior orders results in a Hot-Start recommender that outperforms the Cold-Start
recommender but at the cost of having a low customer coverage.

---

[1] The orders in the test set are segmented based on the number of prior orders in the training set of
the corresponding customer.

Fig. 11 Comparison of the
Cold- and Hot-Start
recommender systems
against the confidence level
of the Hot-Start system

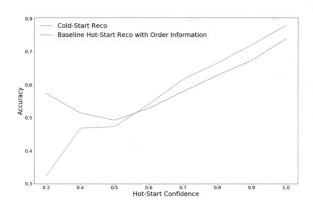

5.4 Hybrid Recommendation Systems

These numbers make a strong case for the incorporation of the Cold-Start recommender system for customers with limited order history. Based on these results, a
hybrid system can be envisioned with each recommender providing a recommendation for the customers it performs best on. One simple yet effective strategy would
switch between recommenders according to the size of the customer's order history. A
more elaborated strategy we experimented with is to use the Hot-Start recommender's
confidence as a hyper parameter, switching to the Cold-Start recommender whenever
the Hot-Start recommender has relatively low confidence in its predictions. Figure 11
presents the accuracy of the two recommenders plotted against the confidence of the
Hot-Start recommender system. To achieve this we segment orders according to the
confidence of the Hot-start recommender and plot the performance of both systems
on those segments. The Hot-Start recommender outperforms its Cold-Start counter-
part in those cases where it is very confident, on the contrary when it exhibits low
confidence (e.g.. <0.5) the Cold-Start recommender proves to be more reliable. Note
that this confidence-based hybrid approach, with an overall performance accuracy
of 59%, is more effective than using the size of the order history (overall accuracy
of 58%, see Hybrid (Orders) and Hybrid (Confidence) in Table 3).

5.5 Minimizing Customers' Cognitive Load

A crucial aspect of any user-in-the-loop recommendation system is the amount of
cognitive load it burdens the customers with. Thus beyond the performance of the
system with respect to accuracy, in practice it is of major importance to minimize
the customer's cognitive load when they interact with the platform. The cold-start
recommender presented in the previous section makes use of 20 explicit customer
data points such as weight, height, etc. which is in reality too high. In what follows, we

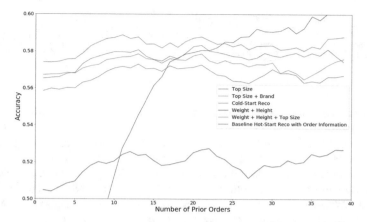

Fig. 12 Comparison of the Cold-Start and the Hot-Start systems. On the x-axis are the number of prior orders (specifically the number of prior orders of each customer in the training set)

deep dive and investigate what performance can be obtained using a small subset of attributes towards providing a low cognitive load and critically reducing the intimate data requirements from the customers on their body shapes, etc. As mentioned in Sect. 4, one of the added benefits of using Gradient Boosted Trees is that they result in an interpretable model, which allows us to estimate the Gini importance of each individual feature in the resulting ensemble. In turn, the attributes that come out as key are Top Size, Weight and Height.

Given the importance assigned to the Top Size attribute by the ensemble, the obvious question that arises is whether the customer provided top size would suffice to predict the size bought by the customer themselves in any future orders. We therefore cross validated this explicit customer information with the sizes a customer actually buys on the fashion platform. We found that in fact customers only buy the size they provided in the questionnaire in roughly 57% of all orders. This observation runs counter to the intuition that customers should be good predictors of their own sizes and highlights one of the many complexities of the size recommendation problem. As customers themselves are only 57% likely to order in their provided sizes, this leaves a remaining 43% of orders where customers are unsure of what size to order and would require accurate support in the form of a size advice.

Figure 12 shows that using solely the size provided by the customer leads, as discussed above, to an under-performing recommender system (marked "Top Size" in the plot). We added a minimum information to it, in particular the brand information, which in turn enables us to use the brand offsets. As can be seen this results in a significantly better recommender system (marked "Top Size + Brand" in the plot) and highlights the importance of exploiting brand offsets when making a recommendation. As expected, the Cold-Start recommender system which has access to the full questionnaire outperforms both these systems.

Table 2 Production results in different countries

Country	Germany	Austria	Switzerland	Netherlands	Belgium	Sweden	Overall
Accuracy	58.36%	57.66%	56.56%	69.50%	68.71%	66.97%	63.90%

Figure 12 also shows the performance of other flavors of the cold-start recommender system using the identified top three attributes: "Weight + Height + Top Size" in purple and "Weight + Height" in red. It is evident that asking only for Weight and Height is not enough to reliably provide a size recommendation. However, we do note that using Weight, Height, and Top Size performs closer to the full Cold-Start recommender than other flavors, and achieves an overall accuracy of 57% instead of 58% (see Cold-Start (W+H+TS) and Cold-Start (All Data) in Table 3).

While "Weight + Height + Top Size" could constitute a good trade-off between high recommender accuracy and low cognitive effort on the side of the customer, however it comes with an underwhelming customer experience which involves having to provide two intimate and privacy sensitive questions on an e-commerce shopping platform. On the other hand "Top Size + Brand" performs closely to that of "Weight + Height + Top Size". This not only highlights the importance of exploiting brand offsets when making a recommendation, but it also comes with the great advantage of not requiring intimate body data from customers. Instead customers are simply asked for their top size in one of their favourite brands. We consider this approach to be the best trade-off between accuracy and customer experience.

5.6 Performance in Production

Following the experiments shown in previous sections and given the encouraging results, we have rolled out our Cold-Start recommender to production in January 2020 on a large e-commerce fashion platform for the Adult Upper Garments category (both Men's and Women's categories). The model is currently live in six countries; Germany, Austria, Switzerland, Netherlands, Belgium, Sweden, serving approximately 3000 orders per day with an overall accuracy of 63.90%. As can be seen in Table 2 accuracy on a per country basis can vary greatly, potentially highlighting the cultural aspect of the size and fit problem which further complicates an already complex problem. This provides an exciting dimension for future work.

Fig. 13 Accuracy of MetalSF (Original [27]), MetalSF (All Data), and MetalSF (W+H+TS) based on the customer predicted size against the number of prior orders in the training set

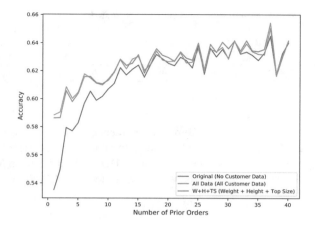

5.7 Leveraging Customer Data for Hot-Start Recommendation

Hot-Start recommender systems, even in the presence of rich order histories, struggle to provide highly accurate size advice. We investigated whether customer data could also be helpful in this case. The Hot-Start baseline [13] cannot readily ingest customer data so we adapted a state-of-the-art Hot-Start deep learning recommender recently proposed in [27] (denoted MetalSF), and show that our findings transfer well to MetalSF. Naturally, customer data helps in the absence of prior purchases, where the accuracy goes from 29% when using the most popular size as prediction to 56% with customer data, on par with our Cold-Start recommender. More interestingly, we plot in Fig. 13 the accuracy against the number of prior orders for various input data: only order history/no customer data (in blue), order history and customer data (in yellow) and order history + Weight + Height + Top Size (in green). Customer data helps significantly up to 10 prior purchases and marginally after 20. Additionally, restricting the customer data to Weight, Height and Top Size has no impact on the performance, indicating that these variables are indeed sufficient to significantly enhance the customers' experience, even where prior purchases are available.

5.8 Summary

The overall performance of the models discussed in the study can be seen in Table 3. Cold-Reco (All Data) refers to the cold-start algorithm that uses all the customer data available and Hot-Reco (Baseline [13]) to the Bayesian Hot-Start recommender baseline. Cold-Reco (W+H+TS) refers to the cold-start algorithm using Weight Height and Top Size only, Hybrid (orders) to the Hybrid recommender based on the num-

Table 3 Results of various approaches studied

Cold-Reco (All Data)	Hot-Reco [13]	Cold-Reco (W+H+TS)	Hybrid (Orders)	Hybrid (Confidence)	MetalSF [27]	MetalSF (All Data)	MetalSF (W+H+TS)
57.89%	53.64%	57.15%	58.41%	59.27%	57.51%	61.45%	61.49%

ber of prior orders, Hybrid (Confidence) to the Hybrid recommender based on the confidence of the Hot-Start recommender. Finally, MetalSF (Original [27]) refers to MetalSF without customer data, MetalSF (All Data) to MetalSF with all customer features, and MetalSF (W+H+TS) to the use of Weight, Height, and Top Size.

6 Conclusion

We addressed a major challenge for the online fashion industry, that of user-in-the-loop personalized size recommendations with low cognitive load. With the aim of building an accurate recommender system that requires only a minimum set of explicit customer data, we further investigated 20 different customer attributes such as weight, height, top size, tummy shape, etc. for the task at hand. We experimented with different versions of the cold-start recommender system and benchmarked them against the state-of-the-art recommender systems with privileged access to rich customer order history. We presented a deep dive on the trade-off between a recommender's performance and a customer's cognitive load, and proposed a solution capable of providing accurate size advice for thousands of new and existing customers with bare minimum customer data needs. Finally we presented our results in a production environment covering six European countries and demonstrated that our approach scales up to large-scale production requirements, performs in practice at the level predicted by the experiments presented here, and to the level of industrial requirements.

References

1. Size charts https://www.adidas.com.sg/help-topics-size_charts. Accessed 13 Jan 2020
2. One size fits none. https://time.com/how-to-fix-vanity-sizing. Accessed 28 Jan 2020
3. Y Yuan, Huh J-H (2018) Cloth size coding and size recommendation system applicable for personal size automatic extraction and cloth shopping mall. MUE/FutureTech 725–731
4. THE ICONIC. https://www.theiconic.com. The customer based size recommendations are accessible on product detail pages. Accessed 11 2019
5. ASOS. https://www.asos.com. The customer based size recommendations are accessible on product detail pages. Accessed 11 2019
6. ABOUT YOU (2019) https://corporate.aboutyou.de/en/. The customer based size recommendations are accessible on product detail pages. Accessed 11 2019

7. Thalmann Nadia, Kevelham Bart, Volino Pascal, Kasap Mustafa, Lyard Etienne (2011) 3d web-based virtual try on of physically simulated clothes. Comput-Aid Des Appl 8:01

8. Surville J, Moncoutie T (2013) 3d virtual try-on: The avatar at center stage. In: Proceedings of 4th international conference on 3d body scanning technologies, 2013

9. Peng F, Al-Sayegh M (2014) Personalised size recommendation for online fashion. In: Proceedings of the 6th international conference on mass customization and personalization in central Europe, 2014

10. Januszkiewicz M, Parker C, Hayes S, Gill S (2017) Online virtual fit is not yet fit for purpose: An analysis of fashion e-commerce interfaces. In: Proceedings of the 8th international conference and exhibition on 3D Body scanning and processing technologies, pp 210–217, 10 2017

11. Sembium V, Rastogi R, Saroop A, Merugu S (2017) Recommending product sizes to customers. In: Proceedings of the Eleventh ACM Conference on Recommender Systems. ACM, 2017

12. Sembium V, Rastogi R, Tekumalla L, Saroop A (2018) Bayesian models for product size recommendations. In: Proceedings of the 2018 world wide web conference, 2018

13. Guigourès R, Ho YK, Koriagin E, Sheikh AS, Bergmann U, Shirvany R (2018) A hierarchical bayesian model for size recommendation in fashion. In: Proceedings of the 12th ACM conference on recommender systems. ACM, 2018

14. G. Mohammed Abdulla and Sumit Borar. Size recommendation system for fashion e-commerce. In *KDD Workshop on Machine Learning Meets Fashion*, 2017

15. Dogani K, Tomassetti M, Vargas S, Chamberlain BP, De Cnudde S (2019) Learning embeddings for product size recommendations. In: Proceedings of the SIGIR 2019 workshop on eCommerce, 2019

16. Sheikh A-S, Guigourès R, Koriagin E, Ho YK, Shirvany R, Vollgraf R, Bergmann U (2019) A deep learning system for predicting size and fit in fashion e-commerce. In: Proceedings of the 13th ACM conference on recommender systems. ACM, 2019

17. Karessli N, Guigourès R, Shirvany R (2019) Sizenet: Weakly supervised learning of visual size and fit in fashion images. In *IEEE conference on computer vision and pattern recognition workshops, CVPR workshops 2019*

18. Weidner N (2010) Vanity sizing, body image, and purchase behavior: a closer look at the effects of inaccurate garment labeling. PhD thesis, Eastern Michigan University, 2010

19. Shi Y, Larson M, Hanjalic A, Collaborative filtering beyond the user-item matrix: a survey of the state of the art and future challenges. ACM Comput Surv 47(1):3:1–3:45

20. Shuai Z, Lina Y, Aixin S, Yi T (2019) Deep learning based recommender system: a survey and new perspectives. ACM Comput Surv 52(1)

21. Yi X, Hong L, Zhong E, Liu NN, Rajan S (2014) Beyond clicks: Dwell time for personalization. In Proceedings of the 8th ACM conference on recommender systems, RecSys '14, pp 113–120, New York, NY, USA, 2014. ACM

22. Catherine R, Cohen W (2016) Personalized recommendations using knowledge graphs: a probabilistic logic programming approach. In: Proceedings of the 10th ACM conference on recommender systems, RecSys '16, pp 325–332, New York, NY, USA, 2016. ACM

23. Pazzani MJ, Billsus D (2007) Content-based recommendation systems. In: The adaptive web, pp 325–341. Springer

24. Misra R, Wan M, McAuley J (2018) Decomposing fit semantics for product size recommendation in metric spaces. In: Proceedings of the 12th ACM conference on recommender systems, RecSys '18, pp 422–426. ACM, 2018

25. Singh L, Singh S, Arora S, Borar S () One embedding to do them all. CoRR, abs/1906.12120, 2019

26. Schein AI, Popescul A, Ungar LH, Pennock DM (2002) Methods and metrics for cold-start recommendations. In: Proceedings of the 25th international ACM SIGIR conference on research and development in information retrieval, pp 253–260. ACM, 2002

27. Lasserre J, Sheikh AS, Koriagin E, Bergmann U, Vollgraf R, Shirvany R (2020) Meta-learning for size and fit recommendation in fashion. In: Proceedings of the 2020 SIAM international conference on data mining, pp 55–63, 01 2020

28. Pavlakos G, Zhu L, Zhou X, Daniilidis K (2018) Learning to estimate 3D human pose and shape from a single color image. In: Proceedings of the IEEE conference on computer vision and pattern recognition, 2018
29. Bogo F, Kanazawa A, Lassner C, Gehler P, Romero J, Black J (2016) Keep it SMPL: Automatic estimation of 3D human pose and shape from a single image. In: Computer Vision – ECCV 2016, Lecture Notes in Computer Science. Springer International Publishing, October 2016

Attention Gets You the Right Size and Fit in Fashion

Karl Hajjar, Julia Lasserre, Alex Zhao, and Reza Shirvany

Abstract Avoiding returns in e-commerce platforms has become a critical issue in terms of both increasing customer satisfaction and decreasing carbon footprint. In the online fashion industry a very large part of the returns is due to size and fit issues that arise from the underlying complexities of shoe and garment manufacturing combined with subjective preferences of customers towards what fits them best. In this context, size recommendation systems capable of estimating a customer's size in thousands of available brands and categories ahead of purchase time are deemed invaluable in dramatically reducing the number of returns related to size and fit. We present a flexible and scalable size recommendation approach that overcomes some limitations of current state-of-the-art work by building upon recent advances in natural language processing and casting the size recommendation problem as a kind of "translation" problem (from articles to sizes) using an attention-based deep learning model for size and fit prediction. Through extensive experimental results, over millions of customers and articles, we demonstrate how this approach is capable of dealing with multiple customers buying from a single account, leveraging cross-category and temporal information to make better predictions, and providing explanations on the final size predictions it produces, thereby helping reduce the potential emotional costs of such predictions for customers.

Alex Zhao—Work done while an intern at Zalando SE.

K. Hajjar (✉) · J. Lasserre · R. Shirvany (✉)
Zalando SE, Mühlenstr. 25, 10243 Berlin, Germany
e-mail: karl.hajjar@zalando.de

R. Shirvany
e-mail: reza.shirvany@zalando.de

J. Lasserre
e-mail: julia.lasserre@zalando.de

A. Zhao
Computer Science Division UC Berkeley, CA 94720, USA
e-mail: axyzhao@berkeley.edu

© The Author(s), under exclusive license to Springer Nature Switzerland AG 2021
N. Dokoohaki et al. (eds.), *Recommender Systems in Fashion and Retail*,
Lecture Notes in Electrical Engineering 734,
https://doi.org/10.1007/978-3-030-66103-8_5

1   Introduction

When shopping for fashion online, customers need to purchase garments and shoes without trying them on to see and feel how they fit. This leads to a great deal of uncertainty in the buying process and to the hurdle of returning articles that customers love but do not fit. Thus, many customers either have to return several purchased articles or remain reluctant to engage in the purchase process altogether. Being able to accurately predict sizes can therefore significantly contribute to increasing customer satisfaction and business profitability through reducing returns, which also reduces the carbon footprint of fashion e-commerce platforms. As an increasing number of people use online fashion stores to shop for articles, these platforms try to support their customers better by providing size information and advice in a passive or active form such as: (1) Size tables and aggregated article measurements [1] provided per brand and article category—this approach requires customers to measure different parts of their body and determine the right size themselves for desired articles by cross referencing their measurements with the size tables; (2) Customer-engaging questionnaires, dialogue-like mechanisms or processing textual feedback [2–4]—customers are asked to provide various explicit personal data such as age, height, weight, tummy shape, hips form, body type, favorite brand, usual sizes, fit preference, etc. to receive a size recommendation; (3) Computer vision and 3D approaches [3, 5–7] providing virtual fit-like solutions based on recent progress in 3D human body estimation [8, 9]—customers are required to provide personal information (as in approach 2.) and/or to take one or multiple pictures of their bodies in tight fitting clothes so a simulated avatar of their body can be built and their measurements predicted to recommend a size; (4) Approaches that leverage existing customers' purchase histories for size recommendation [10–17]—customers with a purchase history automatically receive a size recommendation without any need for providing explicit personal data or images.

All the aforementioned approaches have their own advantages and limitations and the comparison of these radically different methods in tackling the size and fit problem remains out of the scope of this work. In this work we focus on the fourth category of approaches where customers are not required to actively engage in the solution and the size recommender system leverages customers' existing purchase history to provide size recommendations at scale for millions of customers and thousands of brands and articles. Although recommending personalized articles to customers has a long history within machine learning and recommender systems, using methods that can automatically learn from data for size and fit recommendation has only recently received attention [10–17]. What is more, the problem of predicting the right size based on previous purchases is very challenging as: (a) Purchase data is very sparse- a customer only buys a tiny fraction of all the possible articles and sizes that exist; (b) It is also very noisy- a customer can buy various articles for multiple friends and family members in close and neighbouring sizes to their own; (c) The right size for a customer is very subjective- two customers with the exact same purchase history might still buy two different sizes for the same new article based on their perception of the right size; (d) Customers may have a high degree of emotional engagement with the sizing topic—even an accurate size recommendation can come with a high emotional cost for the customer when the recommended size differs from their own expectation.

In this work, we draw inspiration from recent successes of attention-based models in Natural Language Processing (NLP) [18–20] to bring forward a flexible and scalable approach to size and fit recommendation that overcomes the limitations of current state-of-the-art solutions. We propose to model the size prediction problem as an unconventional many-to-one "translation" problem, "translating" from an article to a size given a source sequence of articles (the context). Within this formulation, we take the input source sequence to be a customer's previous purchase history, consisting of all the articles purchased so far, along with the corresponding timestamps and sizes. Then, at a given time, a "query" article is provided to the model and it has to predict (or decode) the correct translation, that is, the right size of this query article for that particular input sequence of articles (which defines the customer in question).

Contributions: The aim of this work is not only to show that the proposed architecture surpasses state-of-the-art performance, but also to highlight the potential and flexibility of a well designed attention-based model in the size and fit problem space. The contributions of our work are as follows:

(1) We demonstrate, for the first time to the best of our knowledge, the value of attention-based approaches in tackling existing challenges of personalized size and fit recommendation in fashion e-commerce. We show such approaches are capable of efficiently leveraging scarce, subjective and noisy purchase data to provide accurate size recommendations.

(2) Our proposed approach overcomes several major limitations of current state-of-the-art approaches. It is trained once for all categories altogether and can digest new data online without having to be fine-tuned for new customers or articles. It takes advantage of the contextual aspect of the problem to leverage cross-category correlations, which is necessary when recommending a size for categories that a customer has never bought before.

(3) We show how explicitly paying different amounts of attention to each previous purchase not only enables predicting sizes more accurately than state-of-the-art size recommendation methods but also enables customers to gain valuable insights as to why a particular prediction has been made for them (in single as well as multi-user behind an account scenarios), thereby moving towards reducing the emotional cost of unexpected size recommendations.

(4) We demonstrate how the adaptability of our approach leads to accuracy improvements on the difficult cold-start problem (new fashion category/user in an account/customer) and low number of previous purchases regime.

The outline of the paper is as follows. We present related work in Sect. 2 and our approach in Sect. 3. In Sect. 4, we provide extensive details on the data and the experimental setup used to build a comprehensive set of experiments. In Sect. 5 we present our results alongside both naive and state-of-the-art baselines, and discuss results on public datasets in Sect. 6. We finally draw conclusions and lay out potential future work in Sect. 7.

2 Related Work

There has been growing literature about size and fit recommendation in recent years. In previous work, a size is always predicted by combining (through a dot product or a concatenation for example) a customer representation with an article representation. Those works can be split into two categories arising from a major conceptual difference: (1) Those which reduce the customer representation to a single vector (either by design, by averaging over past purchases, or by using Gaussian modeling, etc.) [10–16], and (2) Those which build flexible customer representations by considering a list of vector representations of a customer's past purchases [17]. Our work belongs to the latter category.

In the first category, a major series of work focuses on predicting if an article in a given size will be either fit, small or large for a given customer using their past purchase history. References [10–13] all suggest estimating the "fitness" (Small, Fit, Large) between a customer representation and the representation of an article in a given size. While [10, 13] use latent factor models coupled with ordinal regression or metric learning, [11, 12] use Bayesian models with the difference that [12] uses a hierarchical structure and allows directly predicting the probability of any size given a customer-article pair, conditionally on the article being kept (good fit) in contrast with [10, 11, 13] which have to predict the fit (good fit, too big, too small) for all possible sizes one by one. All these works require a numerical mapping of sizes to be applied to other fashion categories than shoes. Finding such a mapping can be difficult and is in itself a research topic [21].

In another line of work, [14] and the Product Size Embedding (PSE) model [15] learn article embeddings for each article-size combination and customer embeddings which are then combined to predict which size of an article would be most likely kept by a customer. Reference [14] pre-trains a skip-gram based Word2Vec [22] per fashion category to learn article embeddings whereas [15] learns them end-to-end, and both get a customer embedding by averaging their purchased article embeddings. In [14], the customer embedding and article embedding are concatenated along with additional article and customer features, and a Gradient Boosted Classifier [23] is trained to classify whether a customer will keep a specific size of an article, whereas in [15] inner products between the customer embedding and the embedding of an article in all possible sizes are computed to obtain scores that are then normalized into probabilities using a softmax.

More recently, a Siamese-like neural network architecture SFNet was introduced in [16] that is able to leverage cross-category correlations. The neural network used first encodes separately the customer and the article, then concatenates the two embeddings and feeds them to fully connected layers before predicting the size. The meta-learning approach MetalSF [17] also learns article embeddings along with size embeddings using fully connected layers, combined with a linear projection to map an article from the latent article space to the latent size space. This mapping is learned using linear regression on the customer's embedded past purchases, and a size is predicted by feeding the projected latent size representation of a new article to fully connected layers which output a distribution over sizes after a softmax.

One category, one model: All aforementioned approaches [10–12, 14, 15], except SFNet [16] and MetalSF [17], suffer from separate model training requirements (one model per fashion category) and expert size mapping development for each category. Aside from the computation costs, having to train multiple models separately dramatically reduces the amount of information available as input to each model (since to predict the size for an article in a given fashion category, only the articles from that same category can be considered). By restricting the number of articles considered in the purchase history of a customer to single categories, such approaches forbid leveraging cross-category information that is potentially useful for predicting the right size for a new article.

Multi-user accounts: The aforementioned methods have different strategies for explicitly dealing with multi-user accounts. Some use manually set thresholds on the range of the sizes purchased by a customer [14] while others use more advanced methods such as Gaussian mixture models [15], Dirichlet processes [12] or hierarchical clustering [10]. In contrast, SFNet [16] relies entirely on the customer embedding to somehow incorporate the information from different accounts during training and later use it at test time. MetalSF [17] does it implicitly through the encoding of gender and category.

Unlike [10–12, 14, 15], the model we present in this work does not have to be trained for different gender-category pairs separately. When given a query article for which we wish to predict the size, we let the model decide—using an attention mechanism—which articles in the previous purchase history are relevant to make a size prediction for that specific query article. This allows us to handle implicitly multi-user accounts and cross-category purchase histories within the predictive model without any additional work to identify multiple users, regroup articles per fashion category, or map sizes to numerical values.

In the remainder of this section we take a closer look at [16] and [17], as our model is similarly trained once for all categories and genders, can effectively utilize cross-category information, and uses deep learning to learn article and customer representations, but yet still bears significant conceptual and architectural differences with those works. The proposed approach differentiates notably from [16] and [17] in that it relies heavily on attention mechanisms [18] (self-attention and source attention) which enables very different types of abstractions compared to that of the fully connected layers [16, 17] and the linear regression used in [17]. A major difference between our model and MetalSF [17] on the one hand, and SFNet [16] on the other hand, is that even though in [16] the sequence of past purchases of a customer is taken into account in some part of the model (when learning customer embeddings), SFNet does not have direct access to this sequence when predicting the size for a new article (no direct comparison to other articles of the purchase history is possible), thereby relying solely on the aggregated information and losing the specific information about each individual article purchased by a customer. This distinction is analogous to the one which originally led to the introduction of attention mechanisms [24] in the decoding part of neural machine translation systems using recurrent neural networks, inducing major improvements in machine translation. In the proposed attention-based model, all past purchases of a customer are given as input context when predicting the size for a new article, irrespective of the articles' genders or categories. The model uses this information flexibly to provide predictions that always depend on the context of previous purchases and not on a single embedding that summarizes the whole purchase history of a customer as in [14–16]. This also implies that, in contrast to SFNet [16], our model can ingest any additional purchase at prediction time without any fine-tuning.

We further outline major leaps in the proposed approach compared to the recent MetalSF [17]. In [17], a customer is represented by their history of past purchases, which are embedded *independently* of the context of the other purchases. In contrast, the self-attention mechanism we employ embeds past purchases mutually based on the context of the other purchases in the history of a customer. Additionally, in [17], the order of purchased sizes in time is not leveraged by the model as the past purchases of a customer are defined as a set. In contrast, the attention-based architecture we put forward is inherently designed for sequential problems (e.g. NLP tasks), allowing us to consider the purchases as an ordered sequence and thus to leverage important information on the evolution of customers' size and fit preference over time. Finally, in [17], the size representation of the query article is obtained through a linear combination of the (context-independent) size representations of the previous purchases with weights that can take any value, whereas our model uses a convex combination (linear combination with positive weights which sum to one). In [17], the weights of the linear combination are obtained by solving a linear regression problem from the representations of articles and sizes, whereas the weights of the convex combination in our model are obtained through an attention mechanism computed on article representations (involving multiple projections and non-linearities), which can yield more powerful representations. In the remainder of this work we demonstrate how the aforementioned critical characteristics of our approach play a strong role in advancing the state-of-the-art in a diverse set of real-world scenarios.

3 Proposed Approach

We build on recent advances in attention models and adapt [18] to the problem of size recommendation. A sentence is the sequence of past purchases of a customer C, referred to as **support purchases**, and is annotated with the associated sizes. When C is faced with a new article, referred to as **query article**, their support purchases are used to infer the size they should purchase this new article in. More formally, for a given sample, the sequence of support purchases of a customer C is denoted by $(p_1, p_2, \ldots, p_n)$, where n depends on the sample, and where a purchase

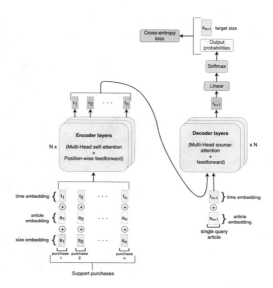

Fig. 1 Attention-based model architecture

$p_i = (T_i, A_i, S_i)$ consists of the timestamp T_i, the article A_i and the size S_i purchased by C. The query article q_{n+1} is defined by (T_{n+1}, A_{n+1}) and consists of time and article information. Note that in practice, many different samples of support/query pairs can be constructed from the list of all purchases of a customer C. Time information is included as input to enable the model to learn how a customer's size and fit preferences or body shape might evolve over time.

The architecture of our model is sketched in Fig. 1. The 3 components are an an encoder, a decoder and a size predictor. (1) The encoder takes as inputs the support purchases of a customer, defined as a sequence of processed vectors including article, time and size information, as described in Sect. 3.1. Each support purchase is encoded into a numerical vector using self-attention to leverage the context of the other purchases. (2) The decoder takes as inputs the sequence of encoded support purchases and a query article, the size of which should be inferred. The query article is a processed vector including article and time information as described in Sect. 3.1. The decoder uses source-attention to leverage the relevant information contained in the encoded support purchases and encodes the query article. (3) The size predictor applies a linear transformation followed by a softmax operation to convert the encoded query article into size probabilities. In practice, the set of available sizes for a particular article is a small subset of all the sizes known to the model and a size mask is used to normalize the size probabilities appropriately (all the mass is given to the sizes available for a given article). The size actually purchased by the customer for that query article is used as **target size** to compute a categorical cross-entropy loss (Fig. 2).

In this paper, a query article is an article actually purchased (and kept) by C, for which we know the timestamp of the purchase and the target size. In contrast, in a real

Fig. 2 Category cold-start sample

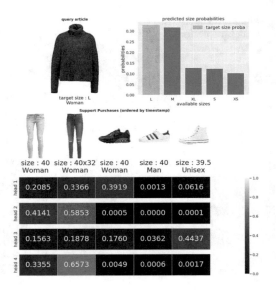

production setting, the query article would be a new article that a customer would be willing to purchase and the timestamp would be the current date. The correct size of this query article for that customer would then be unknown and the model would be asked to produce a recommendation.

3.1 Inputs and Embeddings

This section describes how the raw data is transformed so it can be processed by the encoder and the decoder.

Timestamps. The integer representation T_i of timestamps is given by the number of days between the timestamp itself and a fixed reference date. Note that the reference date is arbitrarily set to a date prior to any purchase in the dataset, but its actual value has no impact on the performance.

Articles. Articles A_i are defined by a set of categorical attributes, namely high-level category (textile, shoes, sports, etc.), gender (men, women, unisex), fashion category (jeans, sweaters, sneakers, etc.), brand, season and supplier. Each attribute is one-hot encoded and the resulting vectors are concatenated to produce the article representation of dimension 4, 621.

Sizes. Sizes S_i are one-hot encoded into a 1, 162 dimensional vector- the list of sizes includes all the different size systems present in the dataset: numeric (38, 40), standard (S, M), fractions (41 1/3, 42.5), confection (36–38, 40–42), etc.

Article, timestamp and size are each embedded independently into a numerical vector. For size and article, simple embedding matrices are used to convert the vector representations into vector embeddings $a_i = W_a A_i$ and $s_i = W_s S_i$. For time, we use the positional encoding made of sines and cosines described in [18], where we have replaced the index of the position of an element in the sequence by the time representation (difference in days to a reference date), clamped to a maximum value $M = 1825$ ($\simeq 5$ years), so that $t_i = f_{pos}(T_i)$. Note that we have used capital letters for the raw representations and lower case letters for the embeddings. All embeddings are of dimension $d = 256$. The matrices W_a and W_s are learned as part of the model's parameters and are, along with the positional encoding f_{pos}, **shared** across support purchases and query articles. As shown in Fig. 1, the representation of a support purchase is the sum of the embeddings for timestamp, article and size. The representation of a query is the sum of the timestamp and article embeddings only.

Keeping the size embeddings separate allows us to map all sizes to a common continuous latent space, the analysis of which is left for future work. Additionally, it allows us to tie the weight of the linear transformation of the size predictor to the size embedding weights, which was shown in [25] to improve performance on language models. Summing up the embeddings is a design choice, concatenating them led to comparable performance. Note that these embeddings are based on generic features such as brand, category, time and not on hashes, so the model can process new articles without fine-tuning. Similarly, a customer is represented by a list of purchases and not by a hash, so new customers can be directly handled.

3.2   Encoder and Decoder Layers

Both the encoder and decoder use $N = 2$ identical layers, with N distinct sets of parameters, stacked on top of one another. Each layer has 2 blocks: for the encoder, 1. a multi-head self-attention block with $h=4$ heads, 2. a position-wise feed-forward block; for the decoder: 1. a multi-head source-attention block with $h=4$ heads, 2. a standard feed-forward block (the position is not needed as there is a single query article). The source-attention weights are computed using the encoder representations of the support purchases as both key and values following the scaled dot-product attention used in [18]. The feed-forward blocks are composed of a 2-layer neural network with a hidden layer of dimension $d_{ff} = 512$ and GELUs activation [26], as used in [19]. All blocks have output dimension d and are followed by a residual connection [27] and layer normalization [28]. The depth shown in Fig. 1 shows the number of attention heads within a layer, not the N layers.

Let us denote, for each support purchase p_i, $x_i = t_i + a_i + s_i$ the sum of the three embeddings. Then, denoting $f_{\theta_e}^{enc}$ the encoder function with parameters θ_e, the encoder takes inputs $(x_1, \ldots, x_n)$ and produces the sequence of n contextual representations of the support purchases $(r_1, \ldots, r_n)$:

$$(r_1, \ldots, r_n) = f_{\theta_e}^{enc}(x_1, \ldots, x_n) \tag{1}$$

We highlight that our model encodes each support purchase using the context of the other associated support purchases, in contrast to other work where support purchases are unaware of one another and have context-free embeddings.

The output of the decoder is a d-dimensional contextual vector representation of the query article computed using the final source-attention weights (one per support purchase) in the decoder. More formally, denoting by $y_{n+1} = t_{n+1} + a_{n+1}$ the sum of the time and article embeddings of the query article, and by $f_{\theta_d}^{dec}$ the decoder function with parameters θ_d, the contextual representation r_{n+1} of the query article is given by :

$$r_{n+1} = f_{\theta_d}^{dec}(r_1, \ldots, r_n, y_{n+1}) \tag{2}$$

The representation r_{n+1} is linearly transformed into a vector of logits and a softmax gives the probability distribution over sizes. Following [25], the linear transformation has no bias term and is tied to the size embedding matrix. In other words, it is given by $W_s^T r_{n+1}$ where W_s is the size embedding matrix of shape $d \times n_{sizes}$ described in Sect. 3.1.

As shown at the top of Fig. 1, a categorical cross-entropy loss is then computed between the probability distribution over sizes output by the model, and the target size actually purchased by the customer. As mentioned in Sect. 3.1, in practice only few sizes among all possible sizes are available for a particular article. Irrelevant sizes are thus masked out by setting the associated logits to $-\infty$. Following [15–17], masking is done during training **and** at test time.

4 Experimental Setup

This section presents in detail the data, the training pipeline, and the hyperparameters used for the model.

4.1 Large-Scale Anonymized Data

We consider purchase data between 2015 and 2019 from a major fashion e-commerce platform for one European country. Only purchases kept by the customers are considered, the integration of return data is left for future work. This leaves **9M** purchases, **380k** unique customers, **770k** unique articles, **2.2k** different brands and **1,162** unique sizes. This dataset is anonymized and not made public due to various customer privacy challenges and proprietary reasons which lie outside the scope of this work. However the important aspects related to the sizing problem at hand are studied to provide a better understanding of the data. Figure 3b (*resp.* Fig. 3a) shows the distribution of the number of purchases per customer (*resp.* per article) in the dataset, and the corresponding cumulative distribution function. We plotted the distribution only for customers with fewer than 100 purchases, and articles with fewer than 50

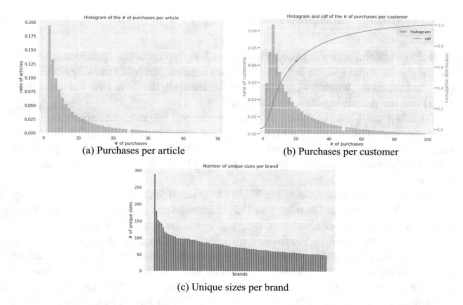

(a) Purchases per article   (b) Purchases per customer

(c) Unique sizes per brand

Fig. 3 Distribution of brands and purchases per customer and article. Left **a**: Histogram of the number of purchases per article for articles with fewer than 50 purchases. Middle **b**: Histogram and cumulative distribution function (cdf) of the number of purchases per customer for customers with fewer than 100 purchases. Right **c**: Number of available unique sizes per brand for the 100 brands with the highest size diversity. Each vertical bar represents one brand

purchases for readability. This represents roughly 97% of all customers and articles in the dataset. We observe that a large majority of articles were purchased less than 5 times, and that more than 60% of the customers have fewer than 20 purchases in their full history of purchases (orange dot on the cumulative distribution function). In those 20 purchases, articles from both genders can be present, and we often have only a few gender-category pairs represented. This means that, for systems which are not able to predict sizes using information from other categories, it would be impossible to serve recommendations in all fashion categories for many customers, or it would take a high number of purchases for a single customer before they can predict a size for most categories, limitations which our model does not suffer from.

Figure 3c shows the number of available unique sizes per brand for the 100 brands with the highest size diversity in the dataset. We observe that nearly all the brands presented provide more than 50 unique sizes for the customers to select from, thereby naturally creating a high degree of uncertainty for the customer regarding which size to purchase even within one brand. On the other hand, even the brand with the highest size diversity only offers a fraction (~25%) of the 1, 162 available unique sizes in the dataset which adds a great deal of complexity for inter-brand size recommendations since different brands might offer different types of sizes for the same fashion category (e.g. numeric vs fractional for shoes).

4.2   Training, Validation and Test Samples

Train/validation/test split. In order to stay as close as possible to a production setting we follow the same split as in [17]. Purchases are ordered by increasing timestamp and the first 80%, denoted $\mathcal{P}_{\text{train}}$, are used for training, the next 10% for validation ($\mathcal{P}_{\text{val}}$) and the remaining 10% for test ($\mathcal{P}_{\text{test}}$).

Training samples. Purchases in $\mathcal{P}_{\text{train}}$ are grouped by customer, giving one sample per customer. Customers with one purchase only are removed and samples are augmented using the procedure below. For each resulting sample, the last purchase provides the query article and its target size, while all other purchases are used as support.

Data augmentation. As noted in Table 1 in column Train (augment.), the training data (only) is augmented similarly to [15]. First we split each customer's purchase history into sequences of maximum length $L = 40$, and consider each sub-sequence as an independent sample. We then re-split each of those samples $n_{\text{splits}} = 5$ times. To achieve that, for a given sample with n purchases, we select uniformly at random n_{splits} integers $k_1, \ldots, k_{n_{\text{splits}}}$ between 2 and n, and then for each of those integers create a new sample by keeping only the first k_i purchases. We keep the original training samples we had before re-splitting and append all the new samples to get the final set of training samples.

Validation samples. Each purchase in $\mathcal{P}_{\text{val}}$ is a query article. The associated support purchases are made of all the purchases in $\mathcal{P}_{\text{train}}$ bought by the same customer.

Test samples. Each purchase in $\mathcal{P}_{\text{test}}$ is a query article q coming from customer c. The associated support purchases are made of subsets of previous purchases from c. Following [17] these subsets vary depending on the chosen scenario.

1. **Offline test scenario**. Standard test scenario where the support purchases are all of c's purchases in $\mathcal{P}_{\text{train}}$.
2. **Online test scenario**. The online scenario simulates a real life production environment where information about a customer is not fixed in time and can be updated with each of their new purchase. Here, the support purchases are all of c's purchases in $\mathcal{P}_{\text{train}}$ plus all of c's purchases in $\mathcal{P}_{\text{test}}$ that were made prior to

Table 1 Upper (a): Number of samples for each split. Lower (b): Model and training hyperparameters

Upper (a)

	Train	Train (augment.)	Val	Test				
# samples	334,170	2,168,017	784,321	815,405				

Lower (b)

	N	h		d	d_{ff}	L	M	n_{splits}	Batch size
values	2	4		256	512	40	1825	5	256

q's timestamp. The first query article associated with c will have only support purchases from the training set, the second query will have a support augmented with the first query article and so on.

3. **Offline (+val)/**4. **Online (+val) test scenarios**. These scenarios mirror scenarios 1 and 2 respectively using $\mathcal{P}_{\text{train}} \cup \mathcal{P}_{\text{val}}$ instead of $\mathcal{P}_{\text{train}}$.

5. **Cold-start online test scenario**. In this last test scenario, we consider customers who only have purchases in $\mathcal{P}_{\text{test}}$ and not in $\mathcal{P}_{\text{train}}$. It mirrors scenario 2 using the empty set $\emptyset$ instead of $\mathcal{P}_{\text{train}}$. That is we start with an empty support sequence for each customer, and update the supports progressively with each new purchase in $\mathcal{P}_{\text{test}}$.

The performance in all of these 5 different test scenarios is summarised in Table 5. Note that the number of samples is the same for the first 4 test scenarios, and that a sample for a given customer has the same query article and target size in all scenarios, only the composition of the support purchases of the sample differs across scenarios. The number of training, validation and test samples are detailed in Table 1a. The cold-start online scenario on the other hand only has 99 k samples as customers who have purchases in $\mathcal{P}_{\text{train}}$ are excluded for this scenario.

4.3 Experimental Details

We detail below different parts of the experimental setup we used to produce the results described in Sect. 5. The code for our Transformer architecture was inspired by the PyTorch code of the Harvard Annotated Transformer[1] [29].

Regularization. We use regularization as in the standard Transformer architecture: additionally to layer normalization, we apply dropout [30] with a rate $p_{\text{drop}} = 0.3$ and label smoothing with a smoothing factor $\varepsilon_{ls} = 0.4$. For a concise description of label smoothing, we refer the reader to Sect. 1.1 of [31]. Smoothing is only applied to the reduced set of available sizes for a particular article, and not to the the set of all possible sizes, allowing us to use a higher value for the smoothing factor. Dropout is applied to the sum of the embeddings before feeding the inputs to the model in the encoder and decoder, and to the output of each sub-block of every layer in both the encoder and the decoder, before the residual connection and the layer normalization.

Optimizer. We use the Adam optimizer [32] with $\beta_1 = 0.9$, $\beta_2 = 0.999$ and $\varepsilon = 10^{-6}$, and vary the learning rate η_k in function of the iteration k just as in [18], using $n_w = 5,000$ warmup steps: $\eta_k = d^{-0.5} \min(n_w^{-1.5} k, \ 1/\sqrt{k})$

Hyperparameters. Table 1b summarizes the values of the model and training hyperparameters. When batching, we pad all support purchases with zeros up to a length of L if needed.

---

[1] https://nlp.seas.harvard.edu/2018/04/03/attention.html.

Hardware and training time. With the values for the hyperparameters described above, we obtain a model with ∼3.7 million parameters. Training this model on the ∼ 2 million augmented training samples on a single NVIDIA Tesla V100 GPU took a little less than 3 days for a total of ∼120 K training steps (15 epochs).

5   Results and Discussion

In this section, we present and discuss the results of different experiments and analyze them following a similar comparative analysis as in [17] to evaluate how our model performs against different criteria, and then draw conclusions about its advantages compared to other work.

5.1   *Overall Performance Comparison*

In this section, we evaluate the model's performance in the offline scenario (defined in Sect. 4.1), which we take as our standard performance comparison scenario, compared to the two simple baselines used in [16] and three state-of-the-art methods [12, 15–17], namely: 1. the popularity baseline (for an article, independently of the customer, the most purchased size for that article is predicted), 2.the Bayesian model presented in [12], 3. the Product Size Embedding (PSE) model [15], 4. the Size and Fit Network (SFNet) [16] and 5. MetalSF [17].

The KDE, Bayesian and PSE methods are trained separately for each category-gender pair. We consider 3 distinct fashion categories: lower-garments, upper-garments and shoes, and two different article genders: male and female. This results in 6 different models for each of these methods. Table 2 shows log-likelihood, top-1-2-3 accuracies and micro-averaged AUC for all approaches on more than 815k test purchases from all categories in the offline scenario. We use the same masking of non-available sizes for **all** methods at test time. The attention-based model performs best with a relatively large improvement compared to [12, 15, 16], and a marginal improvement compared to [17]. We show however in Sects. 5.2.2, 5.4, and 5.5 that the difference in performance increases on the most difficult scenarios. Note that despite the fact that the attention-based model has higher top-1-2-3 accuracies than the other models, it still has lower log-likelihood and micro-AUC than MetalSF [17]. This shows that even though the model predicts sizes more accurately than other models, it is not overly confident in its predictions, which is a typical pitfall of deep learning approaches. This is most likely due to the label smoothing employed in training, which prevents the model from putting too much probability on one single size, making it more robust.

Table 2 Comparison on all categories and size systems in the offline scenario."log lik." stands for log likelihood, "top-k" is top-k accuracy and "mAUC" is micro-averaged AUC

	log lik.	top-1	top-2	top-3	mAUC
Popularity	−2.01	0.29	0.53	0.68	0.69
Bayesian [12]	−1.46	0.47	0.72	0.84	0.79
PSE [15]	−1.47	0.53	0.77	0.87	0.82
SFnet [16]	−1.20	0.55	0.79	0.89	0.85
MetalSF [17]	**−1.04**	0.60	0.83	0.92	**0.89**
Attention-based	−1.11	**0.61**	**0.84**	**0.93**	0.88

5.2 Cross-Category Performance

One advantage of our model, compared with PSE [15], is that it is able to leverage cross-category information to predict sizes. In addition, we show here that our approach performs significantly better than SFNet [16] and MetalSF [17] in exploiting and explaining these cross-category correlations. This advantage can enhance standard size recommendation by using information from other fashion categories than that of the query article, but more importantly can help with the difficult problem of cold-start category recommendation which [12] and [15] cannot tackle. We show below how our attention-based approach is able to deal with both those settings.

5.2.1 Standard Cross-Category Recommendation

We start by showing in Fig. 4a an example where the model attends to different categories to make a prediction in the context where the category of the query article (men's jeans) is also part of the support purchases (rightmost article in the support). Figure 4a is composed of the following. **Top row**: (*left*) query article with its gender and target size, (*right*) model's output probabilities (only the sizes with a probability higher than 10^{-3} among the top 10 sizes are displayed for readability). **Second row**: support articles with their genders and sizes. **Four bottom rows**: weights of each attention head. We observe that the model overall attends mostly to the men's jeans article present in the support, but also pays attention to the men's sweatshirt (3rd article from the right) and the men's t-shirt (2nd article from the right), thereby showing that it does use information from other fashion categories to predict a size.

5.2.2 Category Cold-Start Performance

To quantify the advantage of our approach, we focus here on the category cold-start recommendation problem for upper and lower garments. We show that customers

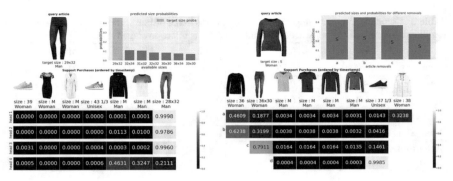

Fig. 4 Size predictions and attention weights. Left **a**: on a standard cross-category sample. Right **b**: when removing different articles from the support

who have never bought in one of these categories can be given a better prediction than with other methods if they have shopped in another category. We compare the performance of our method with the popularity baseline (which the Bayesian, KDE and PSE methods would return since they cannot deal with category cold-start recommendation), SFNet [16] and MetalSF [17], and build two datasets (a and b) which are subsets of the test set used in the standard offline scenario described in Sect. 4.1. For dataset a (*resp.* b) we keep the samples from the offline scenario for which the query article is an upper-garment (*resp.* lower-garment) and the corresponding customer has no upper-garment (*resp.* lower-garment) purchase in $\mathcal{P}_{\text{train}} \cup \mathcal{P}_{\text{val}}$. Results are reported in Table 3a for upper garments and Table 3b for lower garments. The results show that our approach can leverage cross-category information to predict sizes much more accurately than the popularity baseline, SFNet [16] and MetalSF [17]. An example where our model uses information from other categories to predict a size in the cold-start category setting is shown in Fig. 2, which is composed the same way as Fig. 4a.

5.3 Attention Adapts to Changes in the History

We show here how our model adapts its attention when the context (the support purchases) is modified. In Fig. 4b we take an initial purchase history and remove articles to see how the model's attention weights shift. Figure 4b has the same composition as Fig. 4a, except for the following differences. **Top row**: (*right*) predicted sizes for different article removals. **Four bottom rows**: averaged attention weights when removing articles from the support. For each removal, the weights of all 4 heads are averaged at each position in the support to reduce figure size. We remove the following articles from the support one by one (in this order top-down): a. No removal, full set of support purchases, b. the white jacket (rightmost article at index 7), c. the burgundy jumper (leftmost article at index 0), and d. the pair of jeans (at

Table 3 Models comparison on a category of interest for customers who have not bought that category in the training or validation sets

(a) Upper garments (13 k test samples)

	Log lik.	top-1	top-2	top-3	mAUC
Popularity	−1.82	0.31	0.60	0.75	0.64
SFnet [16]	−1.43	0.37	0.62	0.77	0.67
MetalSF [17]	−1.30	0.41	0.69	0.86	0.73
Attention-based	−1.60	0.45	0.73	0.89	0.75

(b) Lower garments (15 k test samples)

Popularity	−2.54	0.24	0.45	0.60	0.71
SFnet [16]	−1.79	0.35	0.57	0.71	0.75
MetalSF [17]	−1.60	0.38	0.61	0.76	0.80
Attention-based	−1.30	0.40	0.64	0.78	0.81

index 1). We observe that more cross-category attention is needed when removing the jumper (leftmost article at index 0) from the support, which is from the same fashion category as the query article. It is also worth noting that when removing the jacket, the model becomes more sure of its prediction. We hypothesize that the jacket might run a bit too small which makes the model slightly more uncertain about the correct size to predict for the query article when it is still part of the support purchases. Other than for this jacket, any article removal from the support makes the model increasingly less sure of its prediction because it has to rely only on cross-category information. It is interesting to observe that, even though the model attends to the jeans with the full set of support purchases, the latter gets roughly twice more attention each time an article is removed from the support. This shows that the model is able to use differently the same piece of information depending on what other type of information is available. Note that the pair of sneakers is actually a unisex article, but from the sizes of the other articles it seems like the model understands that it is an article for the female user behind this account, as it starts paying attention to it when other article are removed, whereas none of the 4 male articles receive any attention even though some are jumpers like the query article.

5.4 Performance on Multi-user Accounts

To evaluate how well our model is able to deal with multi-user accounts we build 6 different experiments keeping subsets of samples from the standard offline setting. 1. (*resp.* 2.) We keep samples where the support has no men's (*resp.* women's) articles and where the query is a men's (*resp.* women's) article (cold-start with no related purchase history). 3. (*resp.* 4.) We keep samples where the support has only men's (*resp.* women's) articles and the query is a men's (*resp.* women's) article (consistent

Table 4 Top-1 accuracy on multi-users accounts (>5k test samples per case)

	Bayesian [12]		PSE [15]		SFnet [16]		MetalSF [17]		**Attention-based**	
	Men	Women	Men	Women	Men	Women	Men	Women	Men	Women
Cold-start (no related history)	0.28	0.28	0.28	0.28	0.32	0.30	0.34	0.30	**0.37**	**0.34**
Consistent (always same gender)	0.44	0.46	0.50	0.51	0.52	0.54	0.55	0.58	**0.59**	**0.60**
Mixed (various genders in history)	0.44	0.47	0.49	0.54	0.48	0.55	**0.55**	**0.61**	0.54	**0.61**

purchase history with always the same gender). 5. (*resp.* 6.) We keep samples where the support has both men and women articles and the query is a men's (*resp.* women's) article (mixed purchase history with various genders). Figure 4a shows a sample from experiment 5, and Figs. 2 and 4b show two samples from experiment 6. In all these samples, we observe that the model correctly attends to past purchases of the same gender as the query article. The results of the experiments are presented in Table 4. For experiments 1 and 2 (no related history), the PSE and Bayesian methods return the baseline prediction. The attention-based approach presents a relatively large improvement over the other approaches [12, 15–17] in most experiments, and the similar performance in the consistent and mixed test cases shows that it is not confused by multiple customers behind an account.

5.5 Online Performance

The attention-based model we propose proves very valuable in an online scenario as it can digest new customers and new purchases after being trained **without** having to be fine-tuned or modified in any way. In contrast, the design of SFNet [16] imposes that: (1) the network has to be fine-tuned to be able to make predictions for any new customer, (2) whenever a customer already present in the database buys a new article and keeps it, the network has again to be fine-tuned to add this new piece of information within the customer embedding and then use it for further recommendations. This is a major advantage of our model for scalability in a practical setting where both the number of customers and the number of purchases they make grow rapidly.

To quantify the advantage of being able to efficiently process purchases online, we compare the performance of the same trained model on the 5 test scenarios described in Sect. 4.1. The results reported in Table 5a are obtained after having trained the model once and fixed all the weights—only the inputs to the model are modified in an online fashion. They show that, as expected, the model is able to leverage additional input information to make more accurate recommendations.

Table 5 Effect of online processing (i.e. updating the customer's support set after each purchase) with attention-based models. (a) Standard online vs offline. (b) Online cold-start performance comparison

(a) Standard online and offline scenarios

Attention-based	Log lik.	top-1	top-2	top-3	mAUC
Offline	−1.11	0.61	0.84	0.93	0.88
Online	**−1.03**	**0.66**	**0.87**	**0.94**	**0.91**
Offline + val.	−1.09	0.63	0.85	0.93	0.89
Online + val.	**−1.03**	**0.67**	**0.88**	**0.95**	**0.91**

*(b) Online **cold-start** (100 k test samples)*

	Log lik.	top-1	top-2	top-3	mAUC
MetalSF [17]	**−1.23**	0.59	0.79	0.88	0.87
Attention-based	−1.34	**0.60**	**0.81**	**0.90**	0.87

In Table Table 5b, we compare the performance of our model to that of MetalSF [17] in the online cold-start scenario where customers (never seen during training or validation) start with empty support purchases, which are updated one purchase at a time. In this scenario, nearly 70% of the samples have less than 4 purchases in the support. The popularity baseline is used for empty support purchases. Our approach is more accurate than MetalSF in this low number of purchases regime. This is probably due to the attention mechanism's ability to quickly adapt to new purchases and predict accurately for categories/genders which are not part of the support purchases, as demonstrated in Table 3 and Table 4, and visualized in Fig. 4b. We leave the analysis of the behaviour on long purchase histories (>40) for future work but hypothesize that even if the linear regression in [17] benefits from more data to learn from, it could still be affected by outliers while the attention model could filter those out if needed.

6 Results on Public Datasets

There are very limited public datasets available for the problem of size and fit, and those (e.g. [4]) mainly focus on leveraging customer metadata for the task at hand to predict "fitness" of an article in a given size. As such, the public datasets introduced in [4] are not directly in the scope of this work as methods evaluated on these use either customer or article hashes or customer metadata such as height, age, weight, body measurements to predict too-small, fit or too-big. This is in contrast to our goal, stated in introduction, of building a model for size prediction based solely on the past purchases of a customer without any need for providing sensitive personal data. However, we consider incorporating additional user metadata within our attention-based approach as a future work avenue, and have thus evaluated the top-1 accuracy, log-likelihood and micro-auc of our method on these datasets. The corresponding

Table 6 Performance comparison on the public datasets ModCloth and RentTheRunway

	Entity embedding		Micro-avg AUC		top-1 accuracy		Log likelihood	
Method/ Dataset	user id	Article id	ModCloth	RentThe-RunWay	ModCloth	RentThe-RunWay	ModCloth	RentThe-RunWay
LF-ML [4]	✓	✓	0.657	0.719	–	–	–	–
SFNet [16]	✓	✓	0.689 ± 0.005	0.749 ± 0.004	**0.690 ± 0.004**	**0.760 ± 0.004**	**−0.758 ± 0.006**	**−0.610 ± 0.008**
SFNet-ne [16]	×	×	0.638 ± 0.007	0.674 ± 0.003	0.683 ± 0.005	0.739 ± 0.002	−0.806 ± 0.009	−0.698 ± 0.006
Attention-based	×	×	**0.818 ± 0.003**	**0.857 ± 0.005**	0.683 ± 0.004	0.728 ± 0.009	−0.850 ± 0.011	−0.779 ± 0.015

results with the comparison to LF-ML [4] and SFNet [16] are shown in Table 6. As in [16], since we do not know the splits used in [4, 16], we used 10 random splits and averaged the results. The performance of our approach is comparable to that of the version of SFNet [16] without any customer nor article embedding, which we refer to as SFNet-ne.

7   Conclusion

s In this work, the use of attention models for tackling the size recommendation problem was shown to address several major challenges of current size recommenders, such as dealing with multiple size systems, cross-categorical and multiple gender recommendations. Additionally, the explainability of the predictions made possible by our approach is a big step towards communicating with customers on the emotionally engaged topic of size recommendations. Our approach surpasses the state-of-the-art in large scale experiments, needs only be trained once for all genders and fashion categories and can easily scale to accommodate new customers and purchases. Future work will focus on studying in depth the embeddings learned by the model in a latent sizing space to extract properties of articles, brands, and customers (from a sizing perspective) as well as on analyzing how integrating pre-trained embeddings learned through another method (e.g. pre-training with BERT [19] or with a size and fit specific method) can enhance the system's performance. We will also study how the flexibility of our model allows incorporating additional customer metadata when it is available, otherwise leaving the presented model unchanged when it is not.

References

1. Size charts. https://www.adidas.com.sg/help-topics-size_charts.html. Accessed September 2020
2. Yuan Y, Huh J-H (2019) Cloth size coding and size recommendation system applicable for personal size automatic extraction and cloth shopping mall: MUE/FutureTech 2018, pp 725–731. 01 2019
3. Januszkiewicz M, Parker C, Hayes S, Gill S (2017) Online virtual fit is not yet fit for purpose: An analysis of fashion e-commerce interfaces. pp 210–217, 10 2017
4. Baier S (2019) Analyzing customer feedback for product fit prediction. 08 2019
5. Nadia T, Bart K, Pascal V, Mustafa K, Etienne L, 3d web-based virtual try on of physically simulated clothes. Comput-Aid Des Appl 8:01
6. Surville J, Moncoutie T (2013) 3d virtual try-on: The avatar at center stage
7. Peng F, Al-Sayegh M (2014) Personalised size recommendation for online fashion
8. Bogo F, Kanazawa A, Lassner C, Gehler PV, Romero J, Black MJ (2016) Keep it SMPL: automatic estimation of 3d human pose and shape from a single image. CoRR, abs/1607.08128, 2016
9. Pavlakos G, Zhu L, Zhou X, Daniilidis K (2018) Learning to estimate 3d human pose and shape from a single color image. CoRR, abs/1805.04092, 2018
10. Sembium V, Rastogi R, Saroop A, Merugu S (2017) Recommending product sizes to customers. In: Proceedings of the eleventh ACM conference on recommender systems, pp 243–250. ACM, 2017
11. Sembium V, Rastogi R, Tekumalla L, Saroop A (2018) Bayesian models for product size recommendations. In: Proceedings of the 2018 world wide web conference, WWW '18, pp 679–687, 2018
12. Guigourès R, Ho YK, Koriagin E, Sheikh A-S, Bergmann U, Shirvany R (2018) A hierarchical bayesian model for size recommendation in fashion. pp 392–396, 09 2018
13. Misra R, Wan M, McAuley J (2018) Decomposing fit semantics for product size recommendation in metric spaces. 10 2018
14. Mohammed Abdulla G, Borar S (2017) Size recommendation system for fashion e-commerce. In: KDD workshop on machine learning meets fashion, 2017
15. Kallirroi D, Matteo T, De Cnudde Sofie, Saùl V, Ben C (2019) Learning embeddings for product size recommendations. In SIGIR eCom, Paris, France
16. Sheikh A-S, Guigourès R, Koriagin E, Ho YK, Shirvany R, Vollgraf R, Bergmann U. A deep learning system for predicting size and fit in fashion e-commerce. In: Proceedings of the 13th ACM conference on recommender systems, pp 110–118. ACM, 2019
17. Lasserre J, Sheikh AS, Koriagin E, Bergmann U, Vollgraf R, Shirvany R (2020) Meta-learning for size and fit recommendation in fashion. In: Proceedings of the 2020 SIAM international conference on data mining, pp 55–63, 01 2020
18. Vaswani A, Shazeer N, Parmar N, Uszkoreit J, Jones L, Gomez AN, Kaiser Ł, Polosukhin I (2017) Attention is all you need. In: Advances in neural information processing systems, pp 5998–6008, 2017
19. Devlin J, Chang M-W, Lee K, Toutanova K (2018) Bert: Pre-training of deep bidirectional transformers for language understanding. arXiv preprint arXiv:1810.04805, 2018
20. Radford A, Jeff W (2019) Rewon Child. Dario Amodei, and Ilya Sutskever. Language models are unsupervised multitask learners, David Luan
21. Du ESJ, Liu C, Wayne DH (2019) Automated fashion size normalization. ArXiv, abs/1908.09980, 2019
22. Mikolov T, Sutskever I, Chen K, Corrado GS, Dean J (2013) Distributed representations of words and phrases and their compositionality. In Burges CJC, Bottou L, Welling M, Ghahramani Z, Weinberger KQ (eds) Advances in neural information processing systems 26, pp 3111–3119. Curran Associates, Inc., 2013
23. Friedman JH (2000) Greedy function approximation: A gradient boosting machine. Annals Stat 29:1189–1232

24. Bahdanau D, Cho K, Bengio Y (2014) Neural machine translation by jointly learning to align and translate, 2014. cite arxiv:1409.0473Comment: Accepted at ICLR 2015 as oral presentation
25. Press O, Wolf L (2016) Using the output embedding to improve language models. CoRR, abs/1608.05859, 2016
26. Hendrycks D, Gimpel K (2016) Bridging nonlinearities and stochastic regularizers with gaussian error linear units. CoRR, abs/1606.08415, 2016
27. He K, Zhang X, Ren S, Sun J (215) Deep residual learning for image recognition. CoRR, abs/1512.03385, 2015
28. Ba J, Kiros JR, Hinton GE (2016) Layer normalization. *ArXiv*, abs/1607.06450, 2016
29. Klein G, Kim Y, Deng Y, Senellart J, Rush AM (2017) Opennmt: Open-source toolkit for neural machine translation. In Proc, ACL
30. Srivastava N, Hinton G, Krizhevsky A, Sutskever I, Salakhutdinov R (2014) Dropout: A simple way to prevent neural networks from overfitting. J Mach Learn Res 15:1929–1958
31. Müller R, Kornblith S, Hinton GE (2019) When does label smoothing help? CoRR, abs/1906.02629, 2019
32. Kingma DP Ba J (2015) Adam: A method for stochastic optimization, 2014. cite arxiv:1412.6980Comment: Published as a conference paper at the 3rd International Conference for Learning Representations, San Diego, 2015

Combining Fashion

The Ensemble-Building Challenge for Fashion Recommendation: Investigation of In-Home Practices and Assessment of Garment Combinations

Jingwen Zhang, Loren Terveen, and Lucy E. Dunne

Abstract Fashion is a domain that poses new and interesting challenges for recommender systems. While most recommendation problems seek a single-point solution (e.g. a product the user will purchase), individual garments must function within a wardrobe system, and must ultimately be matched with other garments to build an outfit. The outfit-building challenge is poorly understood in academic literature and professional practice. Here, we present data from two sources: subjective self-reports from consumers about their outfit-building practices, and assessments (by expert and crowd-sourced assessors) of computer-generated outfit combinations pulled from a real-world wardrobe. Results illuminate the objectives and obstacles of consumers in the daily dressing decision, and support the complexity of building combinations from a large set of individual garments.

1 Introduction

"What am I going to wear?" is a question faced daily by a large portion of the world's population. For some, the answer is simple and direct, and for others it requires a complex, resource-constrained decision-making process. Guy, Green, and Banim [3] describe this process as the "wardrobe moment", a daily mini-crisis in which the individual's wardrobe management techniques are put into play in a time-restricted problem-solving challenge.

The motivations that influence consumption patterns and garment choices in shopping contexts are well-characterized in the clothing and retail literature [10, 15] but little is known about the functional management of those garments at home or the influence of system management on consumption. Management of such a complex

J. Zhang · L. Terveen
Department of Computer Science and Engineering, University of Minnesota, Minneapolis, MN 55455, USA

L. E. Dunne (✉)
Department of Design, Housing, and Apparel, University of Minnesota, St Paul, MN 55108, USA
e-mail: ldunne@umn.edu

© The Author(s), under exclusive license to Springer Nature Switzerland AG 2021
N. Dokoohaki et al. (eds.), *Recommender Systems in Fashion and Retail*,
Lecture Notes in Electrical Engineering 734,
https://doi.org/10.1007/978-3-030-66103-8_6

system is a non-trivial task involving many inter-related variables [7]. Because of the many constraints on cognitive resources during the wardrobe moment, users are likely to employ decision-making heuristics such as satisficing strategies (ceasing search at a "good enough" solution) or availability heuristics and be influenced by cue order (prioritization of garments seen first.) They are also likely to be subject to the size constraints of working memory (generally agreed to be seven ± two items [8]). These factors lead to the reinforcement of a smaller number of items that may consequently become disproportionately utilized, while other garments fall into disuse. This is supported by our prior work with small numbers of participants [2, 13], which show as little as 5% of the wardrobe in regular use.

The challenge of assembling an outfit from component garments is relatively under-studied. Approaches like collaborative filtering based on user and garment attributes [5, 6], classification of "good" (human-generated) from "bad" (artificially created) outfits [11], and discovery of item compatibility across types [12] have been implemented toward the goal of developing methods of effectively building ensembles. However, these approaches often represent an outfit by only one top and one bottom (neglecting more complex outfits) and are based collections of garments from online marketplaces or social networks rather than actual wardrobes. The home wardrobe is a collection curated for the most part by one individual, which may offer some coherence to the components. However, while outfit-building from online repositories may be an exercise in creative exploration (assembling an interesting whole from relatively unlimited resources), managing the home wardrobe is far more resource-constrained, and relies more heavily on combination and re-combination of existing elements. Understanding the parameters and scope of that task is vital to the success of an in-home wardrobe decision-making assistant. Here, we explore the in-home problem from two angles: first, by investigating the challenges, values, and strategies of individual decision-makers in order to better understand how individuals are currently experiencing and managing the complexity of the dressing decision. Second, we form outfits from permutations of garments in a real individual wardrobe, and assess the resulting outfits for wearability in order to better understand the full scope of the decision's complexity.

2   In-Home Outfit Building Strategies

2.1   Methods

To explore the strategies currently used by individuals to choose clothing daily, we conducted an online survey of 194 respondents which is the basis of the bulk of the results presented here. The survey was conducted with participants recruited from Amazon Mechanical Turk, filtered for location (USA only). Respondents were paid $0.50 each for their participation. Participants ranged in age from 18 to 63, 128 were female, and 66 were male. Participants in all instruments used here were assigned a score indicating their position on the consumer spectrum according to a distribution originally described by Rodgers [9] that identifies 5 groups of consumers: fashion innovators, fashion opinion leaders, mass-market consumers, late fashion followers, and fashion isolates and laggards. Consumers on the innovative end of the scale are more likely to make choices based on a desire to differentiate themselves from others, while consumers on the lagging end of the scale are more likely to seek conformity with their social group [1]. To calculate consumer spectrum score, we used a set of nine five-point Likert scale questions derived from [1, 4], assigning points on an inverse scale for self-reported behaviors relating to level of creativity in dress, adoption of new fashion trends, and influence on others' fashion consumption behaviors, for a total spectrum of possible scores from nine (laggard) to 45 (innovator). The consumer-spectrum distributions of our survey participants are depicted in Fig. 1.

The survey covered variables related to the wardrobe moment in five categories, as developed using the preliminary pilot surveys and interviews. The objective of this survey was to investigate the critical variables of dressing and the wardrobe moment that had emerged from our preliminary work. The five categories of questions included were: (1) Perceived wardrobe size and use; (2) Values and objectives in dressing; (3) Constraints of the wardrobe moment; (4) Variables at play in the dressing decision; (5) Outfit-building strategies.

2.2 Results and Discussion

2.2.1 The Wardrobe System

The survey respondents were asked to estimate the size of their "working wardrobe", defined as "the overall number of garments you would wear to work/school or your regular daytime activity". Participants were directed to include only tops, bottoms, dresses, and jackets in their estimates, excluding hosiery, undergarments, outerwear, accessories, etc. Participants who did not have a regular daytime activity or who wore a uniform to work were asked to discontinue the survey. Participants reported average working wardrobe sizes of 25.49 garments (male, SD = 22.08) and 36.72 garments (female, SD = 40.78). When asked to estimate more specifically by reporting numbers of garments in sub-categories (tops, bottoms, and dresses), these averages rose slightly to 27.89 (SD = 21.05) for male participants and 42.10 (SD = 42.31) for female participants. We saw no strong relationships between wardrobe size and consumer spectrum score for men or for women. This finding contrasts with earlier studies such as Workman and Johnson [14] who found significant differences between fashion innovators and fashion followers in need for variety. However, notably such prior work has assessed self-perceptions and desires in dressing, and has not related these theoretical perspectives with empirical assessment of wardrobe contents.

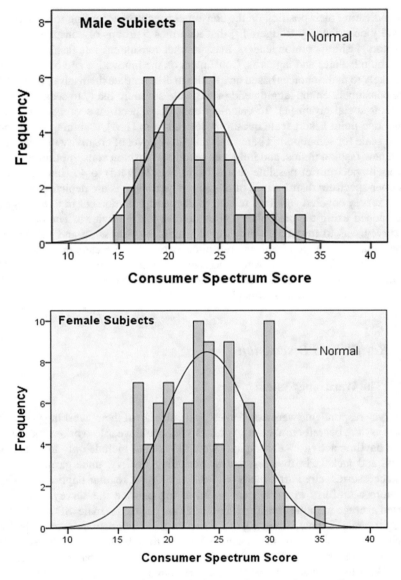

Fig. 1 Distribution of respondents in terms of calculated Consumer Spectrum Score

Survey respondents estimated the percent of their wardrobe in "regular use" (defined as garments worn once per month or more) at 63.00% for female participants (SD = 30.41) and 57.92% for male participants (SD = 34.8).

These results show a vast spectrum of system complexity for the wardrobe across the population. Using a very simplistic combinometrics calculation, a woman's wardrobe of the average size reported by survey respondents (13 bottoms, 24 tops, and five dresses), assuming half of the tops can be worn under the other half, could amount to 1,877 outfits (13 * 12 * 12 + 5). However, some real wardrobes may be far larger and more complex. A wardrobe of the average size reported in [2] (50 bottoms, 159 tops, and 36 dresses) could amount to 316,036 outfits (50 * 79 * 80 + 36).

2.2.2 Objectives in Dressing

Behind the question of the utility of an in-home wardrobe recommender system is the question of what the user aims to achieve in dressing. What implicit values should the system be designed to support? Toward this end, we asked our survey respondents to rate on a five-point Likert scale the level to which 12 value statements were true for them. The results for the statements as described below are summarized in Fig.2.

1. I want to look my best or improve the way I look
2. I want to make use of everything I own and waste as little as possible
3. I want to look trendy and keep up with current fashion
4. I want to have fun putting together outfits, be creative, and express myself
5. I want to fit in and look appropriate every day
6. I want to dress to flatter my body
7. I want to reduce my consumption of clothing
8. I want to be comfortable (physically) in my clothes every day
9. I want to look unique and different in the way I dress
10. I want to get dressed as quickly as possible
11. I want to spend less money on clothing
12. I want to do laundry less often.

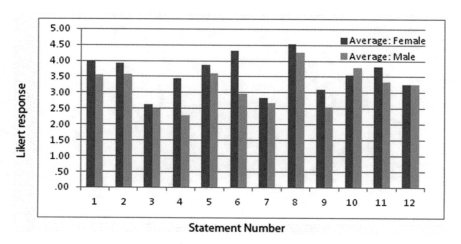

Fig. 2 Male and female participants' ratings of value statements

As Fig. 2 shows, the overall most important objectives for female participants were comfort and dressing to flatter the body. Comfort was the clear leader for men as well, followed by a desire to dress quickly. Of secondary interest for both men and women were looking good and wasting as little as possible in the wardrobe. Fitting in and looking appropriate was of fairly strong interest to both genders, but this factor was stronger than looking good/decreasing waste for men.

Following current trends was of little interest to men or women, and men similarly showed even less interest in having fun/expressing themselves when putting together outfits.

Some factors showed an influence of consumer spectrum score. For instance, the value statement "I want to look unique and different in the way I dress" was more likely to be higher-rated by respondents with higher consumer spectrum scores, as shown in Fig. 3.

A similar effect was seen for both male and female participants in ratings of the "I want to have fun" statement, and the "I want to look better" statement, and for male participants in the "I want to flatter my body" statement. A slight negative relationship was observed for both male and female participants between consumer spectrum score and rating of the statement "I want to get dressed as quickly as possible", as seen in Fig. 4. No visible effect was seen in the other value statements.

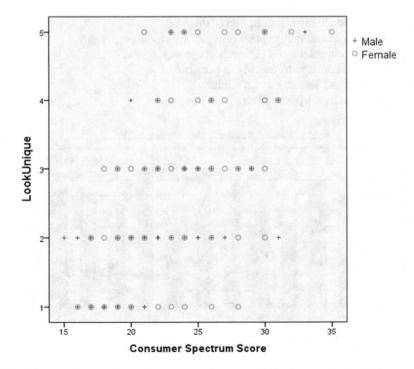

Fig. 3 Influence of consumer spectrum score on importance of uniqueness in dressing

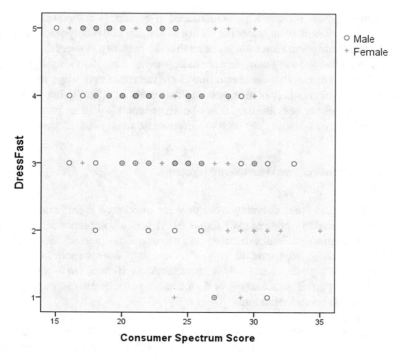

Fig. 4 Influence of consumer spectrum score on importance of speed in dressing

These results imply that there are some more-universal objectives held by users across the consumer spectrum in the wardrobe moment. The objective of being comfortable is perhaps one of the more difficult to capture in a recommender system, given the lack of standardized metrics of fit and haptic preferences in clothing. Characterizing "comfort" is a complex and nuanced task. The more-universal value of looking good, however, implies that many user types would be interested in a decision-support system that augments their aesthetic abilities, or helps prescribe outfits that are likely to be aesthetically successful. An open question for the development of good recommender systems is the precise influences on aesthetic success. Women seem to be more sensitive to the role of body shape in aesthetic success of an outfit than men are, and recommendation systems should likely take body shape into account for women. Male fashion innovators seemed more sensitive to the role of body shape in dressing success, while men across the spectrum were more interested in fitting in and looking appropriate. Women and fashion innovators were more interested in expressive experiences in dressing, at the expense of efficiency in decision-making. Both genders were interested in decreasing waste in the wardrobe. This supports the idea that a recommender system might extend to point-of-purchase decisions, to avoid purchases that are unlikely to provide utility in outfit recommendations.

Interestingly, users were much less interested in trendiness in their dressing decision. Many studies seek to augment the ability to identify and incorporate trends into fashion recommendation. However, it seems that the majority of users are much less interested in trend-following than they are in improving their individual appearance. It is possible, of course, that these two things are intertwined in ways that the user may not be conscious of. It is also possible that users interpret being "trendy" in a negative way (either because they don't perceive their choices as part of a larger trend framework, or because they seek to differentiate from mass-market trends).

2.2.3   Constraints of the Wardrobe Moment

In our studies, time spent deciding what to wear imposed a significant constraint on the decision-making process. We found 61.54% of male respondents reported spending less than five minutes dressing on an average day, and 95.39% spent less than 10 min dressing. We found 42.97% of female respondents spent less than five minutes on an average day, and 80.47% spent less than 10 min. On a "special" day, 38.46% of male participants and 60.94% of female participants reported spending more than 10 min on this decision.

Participants were asked to evaluate on a five-point Likert scale the degree to which or the frequency with which a set of variables increased the difficulty of their daily dressing decision. These variables and the average level of influence reported by male and female participants are shown in Fig. 5. Unlike the set of values outlined in the

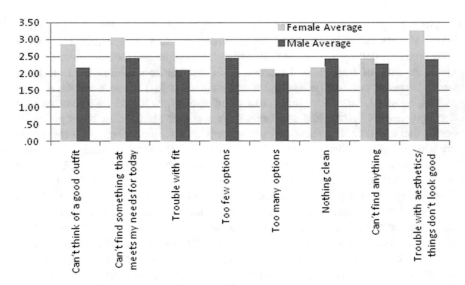

Fig. 5 Male and female participants' ratings of variables increasing the difficulty of the dressing decision

previous section, none of these variables showed a visible relationship to consumer spectrum score.

For female respondents, the most significant sources of dressing difficulty were trouble with aesthetics/looking good, finding an outfit that meets today's needs, and having too few options. Male respondents showed overall lower levels of difficulty arising from these variables, but the most significant sources of difficulty were having too few options, finding an outfit that meets today's needs, and not having anything clean. Having too many options was the least significant source of difficulty for both male and female respondents: an interesting result, considering the evidence discussed above of wardrobe size and percent in use. This may be evidence of boredom with an over-reinforced subset of the population (a result of decision-making heuristics at play) rather than a true lack of options.

Interestingly, the sum of all difficulty ratings for each user (total difficulty) showed no evident relationship to consumer spectrum score. This would indicate that users across the spectrum experience high and low levels of difficulty in dressing. However, the variables influencing difficulty are not likely to be consistent across all users, and must be customized to the individual.

2.2.4 Variables of the Wardrobe Moment

The variables outlined in the previous section describe constraints or sources of difficulty in making the dressing decision: areas where a recommender system might offer assistance to the user. In addition to these constraints, we queried respondents about the variables that influenced the decision of what garments should be worn together and on a given day: variables that will assist the system in generating "good" recommendations. This brings into play temporal variables that are related to the day in question, as well as probing individual priorities in evaluating the goodness of an outfit.

We asked participants to rate on a five-point Likert scale how often or how much a set of variables factored into their daily dressing decision. Those variables and the mean of male and female participants' responses are shown in Fig. 6.

As seen in Fig. 6, the order of relative influence of variables is not exactly the same between male and female respondents: female respondents prioritize the aesthetic element of fit over the comfort component, while male respondents rank the aesthetic component far less influential. For female participants, mood takes precedence over cleanliness, where for male participants, cleanliness and ease of access are far higher in priority. We see similar influence of consumer spectrum score on some variables: fashion leaders are more likely to prioritize overall aesthetic than fashion followers, but the opposite is true for wearing whatever's easiest or seen first, as seen in box plots in Fig. 7.

These data illuminate potential differences in sub-groups of users. Some users may appreciate a pragmatic recommender system that accounts for logistical variables like weather, activity, and laundry status and prioritizes speed of the dressing decision.

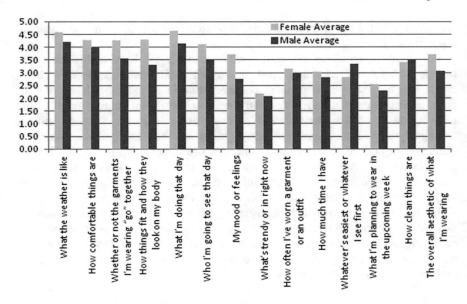

Fig. 6 Male and female participants' ratings of variables influencing the dressing decision

Others may appreciate an exploratory system that affords inspiration rather than prescriptive advice. As might be expected from typical socialization, women seem more sensitive to aesthetic elements that influence coordination of garments and suitability for a given body shape.

2.2.5 Outfit-Building Strategies

Lastly, we sought to investigate the strategies at play when respondents began building an outfit in the wardrobe moment. This category has implications for user interface design. In our preliminary work, which involved more open-ended questions about dressing, many participants reported employing a consistent starting-point in their outfit-building process. Many users cited starting with a single garment (most often the bottom) to build an outfit ("I usually pick out one item I think is particularly appropriate, and build an outfit around it."), but others describe this as a more emotional choice: "I decide which color would match my mood (i.e. I feel….yellow today). I then check the weather to see which items I can combine that have that color and mood and still be practical." As with most of the factors investigated here, in the wardrobe moment these heuristics often help the user to limit the immense quantity of possible options.

Survey participants in this study were asked to indicate which of a set of possible choices (generated from earlier preliminary work) was their most common starting-point in outfit building. As seen in Fig. 8, this is most commonly a top or a bottom

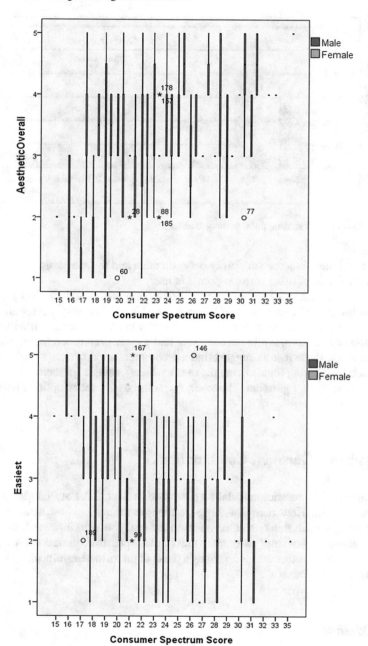

Fig. 7 Influence of consumer spectrum score on importance of overall aesthetic (**a**) and ease of dressing (**b**)

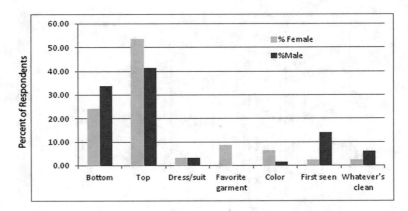

Fig. 8 Most common starting point in outfit building

garment, but there is some variability between male and female participants, as well as some evidence of other starting-points in use.

Again here, male users are more motivated by pragmatics and efficiency (starting with whatever is first seen or clean) than female users, who may start with a color or a favorite garment. Interestingly, male users were more equally likely to start with a top or a bottom, while female users had a clear preference for starting with tops. Together, these results suggest that a recommendation system should start the interaction by offering (or allowing the user to select) a single garment, and building an outfit around that garment. However, for some users, other options may be of interest.

3 Assessing Garment Combinations

The crude combinometrics calculation described in Sect. 2.2.1 at first glance seems improbably high. How many of those garment combinations are actually usable outfits? To approach this question, we used the real-life working wardrobe of one female "fashion innovator" participant and built all possible outfit combinations. A random sample of outfits was assessed by a panel of raters to determine the proportion that is feasibly "wearable".

3.1 Method

Our test wardrobe consisted of 137 items cataloged from a female user's wardrobe: 77 tops, 10 bottoms, 10 dresses, and 5 jackets. Tops were categorized as one of 3 layers (inner, garments that are worn under things; middle, garments that can be worn

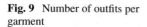

Fig. 9 Number of outfits per garment

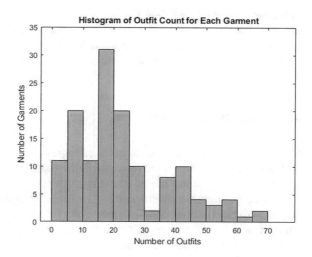

alone or above the inner layer; and outer, garments worn above inner and/or middle garments.)

Each bottom was combined with each inner-layer and middle-layer top, to form the first set of outfits. Then, middle-layer tops were added to inner-layer + bottom outfits. Outer-layer garments were then added to all previous outfits. Finally, dresses were added as single-layer outfits, as well as in combination with each outer-layer top. This algorithm generated 491,185 total outfits. Garments were used in an average of 22.5 outfits, with a distribution between 1 and 70 garments (Fig. 9).

3.1.1 Rating Outfits

A random sample of 1000 outfits was extracted for evaluation from all possible outfits. Each outfit was rated on a 5-point rating scale as follows:

5: This is a great outfit; I can imagine someone looking good wearing exactly this.
4: This is an ok outfit. It might have some style problems, or it might be a little bland, but it's wearable.
3: This is a wearable outfit, but it has some problems. These garments could technically be worn together, but the outfit doesn't work very well.
2: This outfit has serious problems, it would be hard to imagine someone wearing it, but a few people might.
1: This outfit is not wearable; I can't imagine anyone wearing it in public.

The objective of this rating scale was to assess whether or not outfits were "wearable"—and to avoid to the extent possible individual preference, trend, or styling assumptions. Garments were photographed individually, lying flat on a surface (not on a body), and "outfits" were presented as a series of garment photographs. Each outfit was evaluated by 5–7 raters (mean 6.64): 3 "expert" raters (members of the research team, following a calibration exercise) and 2–4 crowdsourced raters. Crowd raters were drawn from 3 sources: our Apparel Design program, targeted advertisements on Facebook, and Amazon Mechanical Turk.

3.2 Results and Discussion

Figure 10 shows a histogram of outfit ratings, separated by rater group (expert = research team, lay = crowdsourced, overall = both).

As seen in Fig. 10 and Table 1, the vast majority of outfits were perceived as "wearable". Expert raters trended toward higher scores than lay raters, which may reflect less of a bias toward personal interpretation of each outfit (a broader perspective on whether or not a set of garments could possibly be successful on some individual). Most remarkable is the difference in "5" score outfits ("I can imagine someone looking good wearing exactly this")—lay raters found only 2.2% of outfits to meet this criteria, versus 34% for expert raters. Interestingly, this may point to the critical "last mile" of fashion recommendation—the gap between a set of garments and those garments being filtered for an individual wearer and styled effectively. When

Fig. 10 Average outfit score by rater group

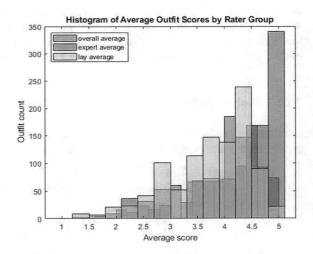

Table 1 Percentages of wearable outfits by rater group

Outfit rating	All raters (%)	Expert raters (%)	Lay raters (%)
>=3 (Could technically be worn together)	90.1	91.6	86.0
>=4 (An "ok" outfit)	61.0	72.5	48.6
=5 (A "great" outfit)	1.2	34.0	2.2

garments are presented without the context of a specific body in 3 dimensions and specific styling choices (shirts tucked in or out, accessories added, etc.) it is unclear how lay users might draw conclusions about the success of an outfit. An expert may be able to imagine a broader range of options and implementations for a set of garments than a lay person, and communicating that expertise is critically important to the resulting dressing decision. This would imply that the best-case recommender system is able to generate an outfit option visualized on a 3D body, complete with styling recommendations. However, this obviously introduces far more complexity into the recommendation and visualization tasks.

It is important to note, however, that while the vast majority of outfits were perceived as wearable, this method is not able to distinguish between perceptually similar outfits: e.g. a sweater worn over several inner-layer tops may present visually the same outfit, but be counted as distinct outfits in this algorithm. Raters did not compare outfits to each other, and would not have looked for repetition in the test set. However, given the size of the set of possible outfits, it is unlikely that much perceptual similarity was present in the test set.

4 Conclusions

The results from our survey of individuals' experiences with the everyday dressing decision highlights consistencies and variation across the consumer spectrum and between individuals in the objectives, obstacles, and strategies employed in the wardrobe moment. Individuals self-report working wardrobe sizes that are relatively small compared to other evidence. Nevertheless, even a small working wardrobe can theoretically produce a vast number of outfit permutations. It is clear that recommender systems can aid users in the efficiency of their dressing decisions and their wardrobe use, and perhaps in improving the aesthetics of their outfit choices. Other objectives like comfort may be harder for recommender systems to influence. Our results also highlight that in some domains, in-home outfit recommendation will require different approaches for different individuals, in order to account for values and priorities as well as sensitivities and preferences: for example, some users may prefer utilitarian systems that prioritize efficiency while others may prefer exploratory systems that afford inspiration and creativity.

The results of our outfit assessment show that both crowdsourced and expert raters find most outfits to be wearable. Even the most conservative assessment—using the 1.2% of outfits that were rated 5 (a "great" outfit) by both expert and crowdsourced raters—yields 5,894 successful outfits from our example wardrobe. However, this result conflicts with user reports of difficulty finding good options and time spent making a selection. Expert and crowdsourced raters in this study evaluated outfits independent of key variables that may determine the ultimate success or failure of an outfit (the variables considered by users in their dressing decision, such as the wearer's physical attributes, contextual elements like weather and activity, and personality aspects that influence aesthetic preferences.) It is clear that there is more work to be done to better understand the underlying factors that predict a successful outfit. Clearly successful outfit recommendation must ultimately account for the wearer and filter the outfit set accordingly. Understanding these relationships (between garments/outfits and the wearer's attributes) is a further challenge for outfit recommendation.

Acknowledgements This work was supported by the University of Minnesota and by the US National Science Foundation under grant #1715200.

References

1. Cholachatpinyo A, Padgett I, Crocker M, Fletcher B (2002) A conceptual model of the fashion process—part 2: An empirical investigation of the micro-subjective level. J Fashion Mark Manag 6(1):24–34
2. Dunne LE, Zhang V, Terveen L (2012) An investigation of contents and use of the home wardrobe. In: Proceedings of the ACM conference on ubiquitous computing
3. Guy A, Green E, Banim M (2003) Through the wardrobe: women's relationships with their clothes. Berg Publishers
4. Hirschman E, Adcock W (1978) An examination of innovative communicators, opinion leaders and innovators for men's fashion apparel. In: Hunt HK (ed) Advances in consumer research. Association for Consumer Research, Ann Arbor, MI, pp 308–314
5. Hu Y, Yi X, Davis LS (2015) Collaborative fashion recommendation: a functional tensor factorization approach. In: Proceedings of the 23rd ACM international conference on multimedia (MM '15), pp 129–138. doi:https://doi.org/10.1145/2733373.2806239
6. Kolstad A, Özgöbek O, Gulla JA, Litlehamar S (2017) Rethinking conventional collaborative filtering for recommending daily fashion outfits. In: RecSysKTL
7. Kwon YH (1987) Daily clothing selection: interrelationships among motivating factors. Cloth Text Res J 5(2):21–27. https://doi.org/10.1177/0887302X8700500204
8. Miller GA (1956) The magical number seven, plus or minus two: Some limits of our capacity for processing information. Psychol Rev 63:81–97
9. Rodgers E (2003) Diffusion of innovations. Free Press, New York, NY
10. Solomon MR, Rabolt N (2008) Consumer behavior in fashion. Prentice Hall
11. Tangseng P, Yamaguchi K, Okatani T (2018) Recommending outfits from personal closet. In: Proceedings of the IEEE winter conference on applications of computer vision (WACV). Lake Tahoe, NV, pp 269–277, https://doi.org/10.1109/wacv.2018.00036
12. Vasileva MI, Plummer BA, Dusad K, Rajpal S, Kumar R., Forsyth D (2018) Learning type-aware embeddings for fashion compatibility. In: Ferrari V, Hebert M, Sminchisescu C, Weiss Y (eds) Computer Vision–ECCV 2018. ECCV 2018. Lecture Notes in Computer Science, vol. 11220. Springer, Cham. https://doi.org/10.1007/978-3-030-01270-0_24
13. Woodward S (2007) Why women wear what they wear. Berg Publishers, Oxford
14. Workman JE, Johnson KKP (1993) Fashion opinion leadership, fashion innovativeness, and need for variety. Cloth Text Res J 11(3):60–64. https://doi.org/10.1177/0887302X9301100309
15. Yurchisin J, Johnson KKP (2010) Fashion and the consumer. Berg Publishers

Outfit Generation and Recommendation—An Experimental Study

Marjan Celikik, Matthias Kirmse, Timo Denk, Pierre Gagliardi, Sahar Mbarek, Duy Pham, and Ana Peleteiro Ramallo

Abstract Over the past years, fashion-related challenges have gained a lot of attention in the research community. Outfit generation and recommendation, i.e., the composition of a set of items of different types (e.g., tops, bottom, shoes, accessories) that go well together, are among the most challenging ones. That is because items have to be both compatible amongst each other and also personalized to match the taste of the customer. Recently there has been a plethora of work targeted at tackling these problems by adopting various techniques and algorithms from the machine learning literature. However, to date, there is no extensive comparison of the performance of the different algorithms for outfit generation and recommendation. In this paper, we close this gap by providing a broad evaluation and comparison of various algorithms, including both personalized and non-personalized approaches, using online, real-world user data from one of Europe's largest fashion stores. We present the adaptations we made to some of those models to make them suitable for personalized outfit generation. Moreover, we provide insights for models that have not yet been evaluated on this task, specifically, GPT, BERT and Seq-to-Seq LSTM.

M. Celikik (✉) · M. Kirmse · T. Denk · P. Gagliardi · S. Mbarek · D. Pham · A. P. Ramallo
Zalando, Berlin, Germany
e-mail: marjan.celikik@zalando.de

M. Kirmse
e-mail: matthias.kirmse@zalando.de

T. Denk
e-mail: timo.denk@zalando.de

P. Gagliardi
e-mail: pierre.gagliardi@zalando.de

S. Mbarek
e-mail: sahar.mbarek@zalando.de

D. Pham
e-mail: duy.pham@zalando.de

A. P. Ramallo
e-mail: ana.peleteiro.ramallo@zalando.de

© The Author(s), under exclusive license to Springer Nature Switzerland AG 2021
N. Dokoohaki et al. (eds.), *Recommender Systems in Fashion and Retail*,
Lecture Notes in Electrical Engineering 734,
https://doi.org/10.1007/978-3-030-66103-8_7

1   Introduction

The role of fashion is constantly growing. In fact, over the last few years it has become one of the world's largest industries, with new trends, products, platforms, and brands constantly appearing. With the vast choice of items available in e-commerce, it has become increasingly difficult for customers to find relevant content, combine it, and match with a specific style.

Search and article recommendations are traditional systems that alleviate this problem. However, many consumers shop new items in order to complement an existing set of garments, or even a full outfit combination. Thus, these customers not only want to be recommended individual items, but a full outfit which is composed of a set of items of different types (e.g., tops, bottom, shoes, accessories), where these items have to be non redundant and visually compatible [5]. For this reason, over the past few years, many stylist-curated services have emerged, that help customers create outfits. However, these human-only-based approaches are not scalable in the growing fashion online market. Further, they may not leverage all the customer information and data that may be available.

Generating and recommending outfits is a huge challenge since it requires the items composing an outfit to be compatible with each other. There are multiple factors that define compatibility or fashion relationship such as brand, cut, color, visual appearance, material, length, and trends. Besides being compatible, the items should be personalized for the specific taste of each customer. Over the past years, a range of work has targeted these problems [3]. Many researchers focused on pairwise compatibility [13, 20], where the outfits are based on item-to-item compatibility. These approaches have the drawback that outfit compatibility is not computed on an outfit as a whole, but on pair-wise article combinations, which also makes them less suited for online serving due to high computation times.

Recently, there has also been work inspired by ideas from the Natural Language Processing (NLP) community, by applying models such as Recurrent Neural Networks (RNNs) [12] to generating full outfits [5]. This has the advantage that the outfit is considered as a whole and not only as pairs of items. However, considering an outfit as an ordered sequence poses unnecessary restrictions. More recently, a new stream of work has used the state-of-the-art model Transformer [19] from NLP, in order to generate personalized outfits [2]. The Transformer-based models BERT [4] and GPT [16] have not been tested on this task yet.

Even though there is a significant effort put into tackling the outfit generation and recommendation problem, to the best of our knowledge, there is no in-depth evaluation and comparison of the performance of different models on this task, including both personalized and non-personalized settings. Moreover, a lot of previous work provides results based only on open-source datasets [10], but not on real-world user data. In this paper we train and evaluate our models using datasets from Zalando,[1] one of the biggest online fashion retailers in Europe, with more than 500k articles and 32M active customers per year.

---

[1] https://zalando.com.

The contributions of our work can be summarized as follows:

- We provide an in-depth evaluation and comparison of different algorithms on the outfit generation task using real-world user data. This includes both personalized and non-personalized approaches. The algorithms are Siamese Networks, Transformer, GPT, BERT, LSTM, and Seq-to-Seq LSTM;
- We adapt the language models BERT, GPT and Seq-to-Seq LSTM to *personalized* outfit generation and extend the Siamese Nets architecture to outfit compatibility.

2 Related Work

Fashion has become one of the world's largest industries. In fact, over the past few years it has gained a lot of attention both in the research community and the industry. Wen-Huang Cheng et al. [3] provide an overview of some of the main applications in the fashion domain, as well as a comprehensive survey of the state-of-the-art research.

Plenty of previous work has focused on pairwise compatibility [13, 20]. To do so, many authors have used Siamese Networks [8], which is a neural architecture that learns an item compatibility function, which in summary computes whether a set of items fit together or not. Veit et al. [21] use them to learn style compatibility across categories, using data from Amazon. Vasileva et al. [18] propose an approach to learning an image embedding that respects an item's type, and jointly learns notions of item similarity and compatibility in an end-to-end model. McAuley et al. [13] use a parameterized distance metric to learn relationships between co-purchased item pairs and used convolutional neural networks (CNNs) for feature extraction. More recently, Polonia et at. [15] leverage Siamese Networks for outfit compatibility, but opposed to previous work, the authors calculate the compatibility score using a fully-connected neural network. However, all these methods do not consider interactions among all the items in an outfit at once.

Inspired by the NLP community, several approaches have been applied to outfit generation. Kuhn et al. [9] propose to use word2vec [14] to learn a latent style embedding for each fashion item solely from the context in which an item appears, by exploiting the curations and expertise of their in-house styling experts. Lee et al. [11] propose Style2Vec, a vector representation for fashion, which learns the representation of a fashion item using other items in matching outfits as context.

The use of RNNs has emerged as an alternative approach to item compatibility. Han et al. [5] use them to model outfit generation as a sequential process. However, considering an outfit as an ordered sequence poses unnecessary restrictions, since permuting the item positions should not alter their compatibility.

The Transformer [19] is a sequence-to-sequence model which has been widely used in NLP. Based on this model, Chen et al. [2] present an industrial-scale Personalized Outfit Generation (POG) model that learns from the user-item and user-outfit interactions and generates a personalized outfit on the fly. Laenen and Moens [10] propose an attention-based fusion method for outfit recommendation which fuses the information in the product image and description to capture the most important, fine-grained product features. Other Transformer-based architectures such as BERT [4] or GPT [16] have also been used to tackle language-oriented tasks. However, there is no work that evaluates these two models on personalized outfit generation.

3 Algorithms

In this section we describe our outfit generation algorithms in detail. For clarity, we divide them into two groups: algorithms for *item compatibility* (we will also refer to those as algorithms for non-personalized outfits) and algorithms for *personalized outfits*. In the first group of algorithms, the learning problem is concerned only with fashion compatibility of a set of fashion items, while the second group of algorithms takes the user preferences into consideration, i.e., the item fashion compatibility is conditioned on the user.

Apart from briefly describing the original architecture that we based our models on, we include the changes we have implemented to some of them (e.g., Siamese Nets and GPT and BERT for item compatibility) in order to be able to adapt them to the outfit generation problem.

We define an *outfit* $x = \{x_1, \ldots, x_n\}$ to be a set of fashion items (garments and/or accessories) with compatible style, where each item can be related to any other item in the set. Depending on the algorithm, we define a user u either as a sequence or a set of past actions (such as add-to-cart, add-to-wishlist, click, etc.), pertaining to items or by questionnaire answers (such as favorite brand, favorite colors, occasion, etc.).

3.1 Item Compatibility

In this section we describe various algorithms for the item compatibility problem, where the task is to learn which items are compatible and could fit together in an outfit. We start by generalizing the Siamese Nets [8] architecture, which we adapt to consider the compatibility of all items in the outfit rather than pairwise compatibility. Further, we describe adaptations of LSTM [6], BERT [22] and GPT [16].

3.1.1 Siamese Nets

The Siamese Nets architecture [1] consists of two identical subnets with shared weights that are inputs to a distance function used for compatibility matching. The distance function can be fixed (e.g., Euclidean) or learned. The identical subnets serve as feature extractors that map the input object into a latent encoding space that represents aspects of the input that are important for compatibility matching. In case of learned similarity, these encodings are concatenated and fed to the similarity block of the network to output the compatibility matching score.

We model the compatibility of two fashion items using a Siamese Nets architecture in a similar way. The identical subnets do not share weights anymore since their respective inputs are different types of objects, for example one for shoes and one for pants. We use sigmoid activation function and binary cross entropy to train the model, where a target of 1 indicates that the items are compatible and 0 otherwise. As positive examples we use stylist-created outfits. Negative examples are obtained by swapping uniformly at random up to m items in a positive example with a random item, where m is the length of the outfit.

Since using this approach models only the pairwise compatibility instead of the outfit as a whole, it has the disadvantage that some items might not be compatible. To this end, we generalize it by adding n parallel subnets, one for each fashion category, for example, shoes, pants, dresses, and jackets. We concatenate the outputs of each of the subnets as described above and include interactions between them, namely the squared Euclidean distance and the Hadamard product to obtain the vector $[x \ y \ (x - y)^2 \ x \cdot y]$. The output of the network is computed in a similar way as before.

It should be noted that unlike the rest of the models described in this chapter that are based on predicting score for all items in the vocabulary, Siamese Nets is a discriminative model and outputs score for a fixed set of items. Hence, it involves computing forward passes on many candidate sets to find one with a high probability of being an outfit. This has the drawback that the architecture is less efficient and it can pose difficulties in online settings where outfit recommendations are generated in real time.

3.1.2 LSTM

The work in [5] considers outfits as sequences instead of sets, where the order of fashion categories is fixed. The authors employ an LSTM [6] to model item compatibility via learning the transitions between items as a proxy. Given a sequence of existing items, a forward LSTM is used to predict the next item in the sequence. Similarly, a backward LSTM is employed to model the previous item in the sequence in order to be able to construct a complete outfit. A zero item is appended to each sequence to serve as a stop token. Given an outfit x, the loss function is given by

$$L\,(x; \Theta) = -\frac{1}{n} \sum_{t=1}^{n} \log \Pr\left(x_t \mid x_1, \ldots, x_{t-1}; \Theta_f\right) - \frac{1}{n} \sum_{t=1}^{n} \log \Pr\left(x_t \mid x_n, \ldots, x_{t+1}; \Theta_b\right), \quad (1)$$

where $\Theta = \begin{bmatrix} \Theta_f & \Theta_b \end{bmatrix}$ denotes the model parameters of the forward and backward model and $\Pr(\cdot)$ is the probability of seeing x_t conditioned on the previous input.

The outfit generation is autoregressive, i.e, the next item is predicted from an initial input set of items which then becomes the next input. To generate outfits with high probabilities, we employ *beam search*, that works by maintaining a set of so-far most likely outfits based on perplexity defined as

$$PP(x; \Theta) = e^{L(x;\Theta)} .$$ (2)

3.1.3 Generative Pre-Training (GPT)

GPT [16] is a popular autoregressive language model based on the Transformer architecture. It adopts the decoder part including the characteristic self-attention mechanism.

Equivalent to the NLP use case, given an outfit x we optimize the following loss function:

$$L(x; \Theta) = \sum_{i}^{n} \log \Pr(x_i | x_1, \ldots, x_{i-1}; \Theta) .$$ (3)

where $\Pr(x_i \mid x_1, \ldots, x_{i-1})$ is the conditional probability of an item x_i given the previous items that is modeled using the Transformer decoder network with parameters Θ.

The main difference to the original GPT language model is that items in outfits do not have an inherent order. Hence, we remove the positional encoding that is added to each token. The outfit sampling at inference time is done in an autogressive fashion similar to the LSTM.

3.1.4 Bidirectional Encoder Representations from Transformers (BERT)

BERT [4] is a masked language model based on the encoder part of the Transformer. It works by pre-training on unlabeled data using two tasks: fill-in-the-blank (FITB) and next sentence prediction. It has been shown empirically that BERT learns rich internal representations during the pre-training phase, which aid fast convergence and high accuracies on different downstream NLP tasks such as named entity recognition. In the following we outline the differences between the original BERT and our adaptation.

We modify the training objective and the output representation. Given an outfit, let $M = x_i$ be the event that item x_i has been masked. The objective function of our BERT model for outfits can be written as

$$L(x; \Theta) = -\frac{1}{n} \sum_{i=1}^{n} \log \Pr(M = x_i \mid x \setminus \{x_i\}; \Theta) ,$$ (4)

where Θ are the model parameters and $\Pr(\cdot)$ is the probability that the model assigns to x_i being the masked item, conditioned on all other items in the outfit.

Similarly to GPT, we have removed the positional encoding of BERT. Furthermore, since there is no equivalent to the next sentence in the fashion domain, we remove the corresponding pre-training task altogether.

3.2 Personalized Outfit Generation Algorithms

In this section we describe algorithms for personalized outfit generation. This problem can be seen as an extension of the item compatibility problem which now includes *context*. This context refers to any information external to the outfit. We distinguish between two main context types, namely *customer actions* (e.g., clicks) and explicit customer preferences in a form of a *questionnaire* (e.g., preferred brands, colors, prices, etc.)

To intuitively understand the advantage of providing context to the model, consider the following example. Suppose a customer has explicitly expressed that she likes casual footwear, such as sneakers. Also, she has previously clicked on items with colorful styles. If we can provide this context to a model it could infer that the customer prefers comfortable, colorful sneakers, and could generate a personalized outfit containing a pair of them.

In the remainder of the section, we start by introducing a generic approach to adapt any algorithm for the item compatibility problem to the personalized outfit generation problem. We then describe how the LSTM-based algorithm from Sect. 3.1.2 can be naturally extended to take the user context into consideration. Afterwards, we describe adaptations of the Transformer, and the BERT and GPT models for personalized outfit generation.

3.2.1 Baseline Algorithm for Outfit Recommendation

Any algorithm applicable to the item compatibility problem can be extended to personalized outfit generation in the following way. First, for each available item in the store, we compute y outfits, where y is sufficiently large to ensure these contain various styles and fashion attributes. Second, given a user u, we rank each outfit with respect to the user item browsing history, for example, by using learning-to-rank or simply by defining a similarity function between a user and an outfit. Such baselines approaches have been already considered in [7]. A particularly effective baseline based on nearest neighbours, defines the similarity between an outfit x and user u as follows

$$\frac{1}{|x|} \sum_{x' \in x} \max_{x'' \in U} \mathrm{sim}\left(x', x''\right) .$$

where $\text{sim}(\cdot)$ defines the similarity between two items, for example cosine similarity between item embeddings. The outfit with the highest score can be chosen as a personalized recommendation.

3.2.2   Sequence-to-sequence LSTM

Sequence-to-sequence LSTMs [17] map an input sequence of arbitrary length to an output sequence of an arbitrary length. This architecture is a straightforward application of the ordinary LSTM cell to general sequence-to-sequence problems. The first LSTM is used to read the input sequence to obtain a fixed-dimensional vector representation of the input. The second one is used to generate an output sequence, conditioned on the state of the first LSTM.

In order to provide personalization, we provide the action sequence of a user u as input to the first LSTM. The output is the outfit considered as a sequence, where the order of fashion categories has been fixed. Hence, the second LSTM learns an "outfit language model" conditioned on the user behavior. The loss and the outfit generation process is similar to that in Sect. 3.1.2 conditioned on u.

3.2.3   Transformer

The Transformer [19] is a powerful transducer model based on *self-attention* with encoder-decoder structure that translates an input sequence to an output sequence. In the Transformer adaptation of the personalized outfit generation problem, proposed in [2], the input to the encoder is the historical user behavior u and the output of the decoder is an outfit x. Each item in the output is generated based on the previous items and the output of the encoder encoding u. Hence, the decoder learns the item compatibility conditioned on the encoded preference signal. The loss function of the Transformer is given by:

$$L(x, u; \Theta) = -\frac{1}{n} \sum_{t=1}^{n} \log \Pr(x_{t+1} | x_1, \ldots, x_t, u; \Theta) .$$ (5)

It should be noted that apart from the user-item sequence that has limited length, the Transformer allows providing more global context, e.g., user segmentation or affinities for brands, styles, colors, etc. This can be done by assigning fixed positions in the encoder reserved for additional contextual embeddings.

3.2.4   Contextual BERT and GPT

The BERT and GPT models for outfits described in Sect. 3.1.4 and 3.1.3 are non-personalized, i.e., regardless of customer preferences or interactions, they generate the same outfits. However, we want to be able to *condition* these models during inference in order to generate better suited outfits for each customer. The context we are using is information about the customer such as season, gender, age, weight, height, preferred brands, preferred colors, and other summarized customer information. We therefore extend both models and make them contextual in the following way.

We embed the context into a vector space which has the same dimensionality as the item embedding. For BERT this context is appended as an additional token and for GPT it is added as a start token. In both cases the models can attend to the context vector and utilize it for prediction. This method resembles the work in [23], where binary information about the sentiment of a sentence is injected into BERT.

While GPT can be naturally used to sample outfits autoregressively, BERT has originally not been designed for generative tasks [4]. Recent works such as [22], however, suggests employing Gibbs sampling to retrieve full-length sentences from BERT. We adopt it by iteratively masking out positions in a randomly initialized outfit and using a trained BERT model to find replacements for them. Note that many[2] forward passes are required for this method, while GPT can generate an outfit of length n in $n + 1$ forward passes.

4 Experiments

In this section we provide offline results and insights on the performance of different algorithms that were introduced in Sect. 3. We first introduce the datasets and the features and then evaluate the non-personalized algorithms followed by the personalized ones.

4.1 Datasets

In this section we present the datasets used to train and evaluate our models. They come from Zalando, a hub for digital fashion content in Europe. In the online shop, customers can purchase or seek for inspiration about garments and style via, for example, outfits.

The hand-crafted outfits we use to train our models are created in three different ways: (1) outfits from content creators (such as stylists) on the website (Shop the Look, STL), (2) influencer-created outfits (Get the Look, GTL),[3] where influencers

---

[2]The exact number of forward passes for sampling a single outfit from BERT depends on the implementation. We found it to be ideally at least an order of magnitude higher than the outfit length.

[3]https://en.zalando.de/get-the-look-women.

Table 1 Comparison of the key properties of Zalando's GTL & STL and Zalon's outfit dataset

Dataset	Personalization	# Outfits	#Articles	Avg. outfit length
Zalando GTL & STL	Click history	251,891	64,748	4.50 articles
Zalon	Questionnaire	380,808	30,619	4.96 articles

assemble their own outfits, and (3) via Zalando's personalized styling service Zalon[4] where stylists create outfits customized for each customer individually.

- **Zalando outfit dataset (GTL and STL):** This dataset consists of around 250k hand-created outfits that have been published on Zalando, containing a total of 1M distinct items. This includes STL styled by Zalando creators, provided as an inspirational supplement to the item on the product detail page, and the influencer-created outfits available on GTL. Each of these outfits is composed of a single item per body part that can be worn together, occasionally accompanied by a fashion accessory.
- **Zalon outfit dataset:** A dataset of around 380k recent outfits, each of which has been handcrafted for a specific customer by a professional stylist. A Zalon stylist assembles an outfit based on *questionnaire* answers that a customer provides, where they express their style preferences, provide body features to the stylist, and specify price expectations. Based on this information, the stylist creates a personalized box consisting of two outfits, where each consists of up to seven articles, for example shoes, pants, t-shirt, sweater, and jacket. There can be multiple articles of one type, for example multiple pants or tops, but all of them are compatible amongst each other. Zalon's dataset contains about 30k distinct articles from the Zalando fashion store. To restrict ourselves to this limited set of distinct articles, we removed the long tail of articles which appear less than eight items.

In Table 1 we show a summary of the two outfit datasets we just described. With the previous datasets, we can solve the compatibility problem. However, in order to be able to cater for the specific taste of our customers, we also make use of customer context. The following two datasets contain user specific data in the form of clicks and questionnaires:

- **Zalando click dataset:** This dataset consists of user click actions (clicks on articles, additions to the wishlist, etc.) on a single item and user actions on whole outfits that are available on Zalando. In total there are close to 1M outfits per year created on the Zalando web page available to approximately 32M active customers. We aggregate the past actions per user over a period of one month and create training samples consisting of outfit interactions together with the preceding item actions such as click and add-to-wishlist. The outfit actions are taken into consideration only if there are at least five item actions preceding it and contain

---

[4]https://zalon.de.

at least four items, excluding accessories. We exclude fashion items that are rare and occur less than three times in the action data. This way, we obtain around 6M valid training samples that contain around 200k distinct outfits and 100k distinct items.

- **Zalon questionnaire dataset**: In Zalon, each customer that requests a personalized outfit needs to fill a detailed questionnaire, which gives information about style preferences, provides body features to the stylist, and specifies price expectations. The total number of features we collect is over 30, with some examples of questionnaire fields being the shoe size, no-go dress types, favorite brands, favorite colors, hair color, body height and weight, and the occasion for which an outfit is needed. The total amount of questionnaires for training our models is around 250k.

Our datasets are proprietary and cannot be released for customer privacy reasons. The Zalon dataset is distinct from other datasets in the domain because of its rich questionnaire features which contain significantly more information about a customer than their click or purchase history. It is therefore especially promising to use in combination with personalized models.

4.2 Item Representation

For all our models, we use the same item representation that contains a 128-dimensional image embedding extracted from the penultimate fully-connected layer of a fine-tuned ResNet-50 CNN computed from the packshot item image. This vector is then concatenated with a vector of learned embeddings of categorical item attributes, in particular: *category, brand, season, color, gender, material* and *pattern*. We use a softmax layer to predict a probability distribution of a subset of the full vocabulary of items appearing in the training and test sets. We keep only the items with frequency larger than a predefined threshold of 8 occurrences.

4.3 Non-Personalized Models

In this section we present the results of our experiments on the non-personalized algorithms. We first describe the implementation details and define the metrics followed by evaluation on both the Zalando and the Zalon outfit datasets introduced in Sect. 4.1.

4.3.1 Implementation Details

- **GPT** and **BERT**: We use four layers with eight attention heads each and set the model dimensionality $d_{\mathrm{model}} = 128$. We use batch size of 512 and a dropout rate of 1%.
- **Siamese Nets**: We use two fully-connected layers for the feature-extractor subnets and two fully-connected layers for the item compatibility part of the network. Each layer has 64 ReLU units. We generate the negative samples by randomly changing from one up to n items in each outfit in the training set, where n is the size of the outfit. We use batch size of 32.
- **LSTM**: We use the setting from [5]: a single-layered LSTM cell with 512 hidden units and a dropout of 0.3. We train we batch size of 64.

We randomly split our data into 90% train and 10% validation data.

4.3.2 Metrics

To evaluate the quality of the non-personalized models, we adopt three well-known metrics:

Perplexity (PP)
The perplexity is a common metric in the NLP domain. It reflects how well the model has learned an underlying distribution in an autoregressive fashion. In our case, a low perplexity indicates that the model is performing well at sequentially generating samples from the approximated outfit distribution. For a single outfit, the PP is defined based on the average cross-entropy (CE) as

$$\mathrm{CE}\,(x; \Theta) = -\frac{1}{n} \sum_{t=1}^{n} \log \Pr\,(x_t \mid x_1, ..., x_{t-1}; \Theta)\,, \tag{6}$$

$$\mathrm{PP}\,(x; \Theta) = \exp\,(\mathrm{CE}\,(x; \Theta))\,, \tag{7}$$

where x is a ground truth outfit. To report the PP for the validation dataset we average it across the outfits. In an effort to make BERT comparable to GPT, we compute the perplexity for BERT by masking every item once and removing the respective context to its right.

Fill In The Blank (FITB)
The FITB recall at rank r, also abbreviated as FITB@r, measures the model's ability to complete an outfit where one item was masked out. It represents the probability that the ground-truth article is among the top r predictions made by the model. In our case, we compute r@1, r@5, r@25, and r@250.

We implement FITB for GPT as follows. First, the masked item x_i is removed from the outfit and the remaining items $x_1, \ldots, x_{i-1}, x_{i+1}, \ldots, x_n$ are fed into the network. The network's prediction at position n is then interpreted as prediction for the masked-out item. Note that this is only reasonable if GPT is trained with randomly shuffled outfit sequences. To clarify this, assume that GPT is trained with outfits which are sorted as follows: shoes, pants and shirts. If the pants are removed and the model is presented with shoes and a shirt to predict the missing pants, this would constitute an out of distribution case since the model has never seen this combination in this order before. On the other hand, if the outfits are shuffled, the model has likely already seen such combinations.

Compatibility Prediction (CP)

The outfit compatibility metric evaluates a model's capability at distinguishing compatible from non-compatible outfits. For each compatible outfit we generate a non-compatible example by replacing one item at a randomly selected position by another random item from the vocabulary. This replacement method yields a new dataset of outfit pairs, where each outfit is labelled as either compatible or non-compatible. The task constitutes a binary classification problem, where we use the area under the curve (AUC) of the receiver operating characteristic (ROC) as the CP metric. To calculate the classification score for BERT, GPT, and the RNN, we compute an outfit probability as $\exp(-CE(x; \Theta))$ and treat it as a classification score for computing AUC.

4.3.3 Results

In Table 2 we present the results for the different non-personalized algorithms using the Zalando-outfits dataset. The Siamese Nets serve as a baseline and we can see that they consistently perform worse than the rest of the models on all metrics (we do not report on perplexity since they are not a language model). We attribute the worse performance to the following two reasons: first, unlike the rest of the models, on prediction Siamese Nets do not produce probability distribution over the entire vocabulary, but rather each item in the vocabulary must be ranked in isolation. Therefore, the scores the model outputs cannot be directly compared to each other and often sub-optimal choices are made by picking the item with highest score. Second, the generation of negative examples needed by the Siamese Nets is also suboptimal since it relies on the strong assumption that randomly changing items in the outfit always results in set of items that are not compatible.

Table 2 Comparison of non-personalized models on the Zalando outfit dataset

Model	PP	CP (%)	FITB@r1(%)	FITB@r5 (%)	FITB@r25 (%)	FITB@r250 (%)
Siamese Nets	–	73.7	0.4	1.3	5.2	23.7
LSTM	34,290	68.6	2.4	5.8	7.9	13.1
GPT	**92**	96.9	17.7	26.9	37.0	52.2
BERT	182,586	**97.9**	**49.3**	**71.7**	**88.2**	**98.6**

Table 3 Comparison of non-personalized models on the Zalon outfit dataset

Model	PP	CP (%)	FITB@r1 (%)	FITB@r5 (%)	FITB@r25 (%)	FITB@r250 (%)
Siamese Nets	–	71.9	0.1	0.2	0.6	4.5
LSTM	28,637	64.1	0.7	1.6	2.9	6.8
GPT	**1,212**	**92.1**	2.4	6.7	15.3	40.8
BERT	9,934	89.0	**4.8**	**12.5**	**26.1**	**61.9**

BERT and GPT show opposite performance on different metrics. While BERT achieves higher accuracy on the FITB metrics, GPT has a much lower (better) perplexity. That can be explained by the fact that BERT is trained on a task similar to the FITB metric, giving the model a significant advantage, while the GPT is trained on a loss that resembles perplexity, hence performing much better on this metric. Although we expected similar performance between GPT and LSTM, we observed consistently worse performance of the LSTM on both of our outfit datasets.

On the compatibility task (CP), BERT obtains the best results with 97.9%, followed very closely by GPT, with 96.9%. The high AUCs are remarkable given that the training dataset does not contain any negative samples and the models were never explicitly trained on the task of distinguishing compatible and non-compatible outfits.

Our results confirm that, similarly to the NLP domain, GPT is much better suited for generation purposes than BERT. On the other hand, BERT excels at completing outfits with a single missing item. Investigating the usefulness of the contextualized internal representations that BERT computes for each item is a topic left as future work.

In Table 3 we present the same results on the Zalon outfit dataset. Comparing the results from the two datasets against each other, we see that while they are consistent with respect to the model performance, they are systematically better on the Zalando outfit dataset. While this deserves further investigation, we believe that the difference may be caused by the different item distribution in influencer and stylist outfits. More specifically, the Zalon stylist outfits are created for a personalized customer and a variety of occasions, which means that they are more diverse. This more heterogeneous distribution might be harder for the models to learn. Furthermore, the Zalon outfits in average contain more items than the Zalando outfits.

4.4 Personalized Outfit Generation

In this section we present and interpret the experimental results of the personalized outfit generation algorithms. We use two different representations of the user context: action sequences (customer click dataset) and a questionnaire answers. We define metrics suitable for outfit recommendation that capture different aspects of the quality of the generated outfits. Finally, we evaluate the algorithms on past click and purchase (kept item) data.

4.4.1 Implementation Details

- **Contextual GPT** and **Contextual BERT**: We use the same architecture as in the non-personalized experiments with the addition of the questionnaire embeddings described in Sect. 3.2.4. Moreover, in order to compare GPT and BERT on our metrics their sampling methods have to be aligned. Since BERT is sampled with the fixed length of the original outfit, we apply the same procedure for GPT. That means, instead of sampling until the stop token is reached, we sample GPT with the fixed length of the original outfit as well.
- **Transformer**: Both the encoder and the decoder consist of two layers, with 12 attention heads each. The model dimensionality d_{model} was selected to be equal to the total size of the input embeddings, namely 216. Each position in the encoder is represented by learned embeddings of the item attributes introduced in the beginning of the section, concatenated with a one-hot representation of the event type (item click, wishlist-change, cart-change) and a normalized scalar value for the action age, counted in number of days between the outfit and the item click. We use dropout of 0.1 and batch size of 64.
- **Personalized Siamese Nets**: We adapt Siamese Nets introduced in Sect. 3.1.1 to include personalization as follows. For each relevant item in the assortment, we precompute up to 100 outfits and calculate the nearest neighbors between the browsing history of the user and each of the precomputed outfits that contain a particular item the user is currently interacting with. The nearest neighbors are calculated based on Eq. 3.2.1 by using cosine similarity between the image embeddings. We show the top-1 outfit with highest similarity to the customer.
- **S2S LSTM**: For the encoder and the decoder of the sequence-to-sequence LSTM we use the same setting from the previous section. We sort the outfits by fashion category and train a forward and a backward model in order to be able to generate full outfits from tip to toe.

We evaluate the action sequence based algorithms on the Zalando click dataset generated from interactions with outfits that complement the main item, which we call anchor, on the product detail page (see Fig. 1). We evaluate the questionnaire-based algorithms on Zalon's order and kept items dataset. We use 10% of 30 days of aggregated action data for evaluation. We use a time-based split, leaving out the last few days of data for evaluation and the rest for training.

Complete the look
Outfit inspiration

312,95 € ~~388,95 €~~
CLEARY - Leather jacket - black
JOOP! Jeans

43,95 € ~~72,95 €~~
Shirt - royal
J.CREW
~40 %   Premium

92,95 € ~~154,95 €~~
COOPER - Jeans Tapered Fit - mid blue
CLOSED
~40 %   Premium

Fig. 1 Algorithmic (right) and stylist (left) outfits side-by-side on the product detail page. The item highlighted in red is the anchor item based on which the Transformer model generated the outfit taking the customer's click history into consideration

4.4.2 Metrics

We use the following set of metrics to asses how diverse the generated outfits are and how well they match individual customer preferences, which is reflected by what the customer has clicked on or purchased.

Fashion attribute click-through rate (CTR)
Many combinations of items could be compatible with a given anchor item since compatibility is defined by multiple factors such as brand, style, color, etc. If an algorithm does not reproduce an exact match, it might be due to the large combination of possible compatible items. Another item might fit perfectly yet be very different visually to what the user has interacted with in the historical click data. To this end, we use proxy metrics for assessing the matches that are based on attributes, in particular the following combinations *brand-category*, *color-category*, *brand-color-category*. For example, if the generated outfit contains an item with the same brand and category as the clicked item, then we consider this a brand-category match. The brand-category hit rate is then the fraction of generated outfits with both, brand and category match.

Fashion attribute keep rate (KR)
Keep rate refers to the fraction of items a user has bought from a shipped full outfit. The proxy metrics for exact KR are defined in the same way as those for CTR, based on matching fashion attributes. These metrics are used in the experiments on the the Zalon dataset.

Personalization rate
A proxy metric to estimate the ability of the algorithm to personalize, i.e., generate different outfits for different users. It is defined as the ratio of distinct outfits o among

Table 4 Personalized action sequence based algorithms evaluated on the Zalando click dataset

Metric	Siamese Nets (%)	Transformer (%)	Seq-to-Seq LSTM (%)
Brand-category CTR	5.8	**40.8**	9.4
Color-category CTR	9.3	**40.2**	12.8
Brand-color-category CTR	2.7	**35.6**	7.4
Personalizaton rate	10.7	24.1	**51.9**
Item diversity rate	7.7	31.4	**35.7**

the outfits recommended to n different users. It would be 100% if every user was served a unique outfit.

Item diversity
It is a desirable property of an algorithm to generate outfits with diverse items. The reason for this is: first, showing outfits with a narrow set of items might hurt customer experience, for example, due to obvious repetition of popular items. Second, the outfits should inspire the customer with a broader assortment of items available at the fashion retailer's catalog. We therefore define item diversity as the ratio between the number of unique and the number of total items used to generate all outfits for all users during the offline evaluation.

4.4.3 Results

Table 4 reports the results of the action-sequence-based algorithms evaluated on the Zalando-click dataset. We use the Siamese Nets algorithm as a baseline since it is widely used in the literature. The Transformer outperforms S2S LSTM and Siamese Nets on all CTR-based metrics. We attribute this to the ability of the model to effectively learn the underlying outfit probability distribution and in the same time learn complex interactions between user click behavior and an outfit the user might be interested in. Moreover, the Transformer and S2S LSTM generate more diverse outfits unlike the Siamese Nets which tends to favor certain items. Finally, the S2S LSTM displays a higher personalization rate albeit significantly lower CTR rate than the Transformer, the main metric for which we optimize. We hypothesize this is due to higher instability of the LSTM in learning the underlying outfit distribution since the LSTM also tends to generate non-valid outfits more often than the Transformer.[5] We leave this further investigation and fine-tuning for future work.

In Table 5 we compare Contextual GPT and Contextual BERT against their non-personalized counterparts. Here we use the non-personalized metrics from Sect. 4.3 except for the compatibility, which we have excluded since personal context should not significantly affect the ability to distinguish compatible and non-compatible out-

---

[5]A random algorithm would result in close to 100% personalization rate.

Table 5 Personalized models compared to their non-personalized counterparts on the Zalon dataset

Model	PP	FITB@1	FITB@5	FITB@25	FITB@250
GPT	1,212	2.4%	6.7%	15.3%	40.8%
Contextual GPT	**728**	3.1%	8.5%	19.8%	49.5%
BERT	9,935	4.9%	12.5%	26.1%	61.9%
Contextual BERT	15,779	**5.9%**	**14.5%**	**30.9%**	**68.1%**

Table 6 Results of the personalized, questionnaire-based algorithms on the Zalon dataset. Metrics are related to purchases; KR stands for keep rate. For example a brand-category KR of 100% would mean that for every item that was kept by a user there is one item with the same brand and category in the personalized, predicted outfit for that very user

Metric	Contextual GPT	Contextual BERT
Brand-category KR	**2.0%**	0.6%
Color-category KR	**2.3%**	1.6%
Brand-color-category KR	**0.7%**	0.2%
Personalization rate	**0.5%**	**0.5%**
Item diversity rate	5.6%	**33.6%**

fits. Regarding the FITB task, we see a significant performance increase for both algorithms: more than 25% for GPT and more than 10% for BERT. This shows that the models makes use of the additional information such as preferred color or brand, to predict an item similar to the stylist's choice, who have incorporated this information into their decision process. Furthermore, the perplexity of Contextual GPT decreased by 40% for the same reasons, however, BERT's perplexity increased. This can be explained as before by BERT being trained on the FITB task. Namely, the better it gets on the FITB task, the worse its model perplexity gets.

In Table 6 we compare Contextual GPT and BERT against each other on the more fine-grained personalization metrics defined above. Here we see that GPT performs in general better, i.e., picks items that are more similar to the items the customer has actually kept. While both models benefit from personalization in terms of FITB, this improvement does not seem to translate proportionally in both models in terms of quality of the generated outfits. This might be caused by the different outfit sampling methods: autoregressive generation is a seemingly more efficient and effective generation method than Gibbs sampling that we employ for BERT. We hypothesize that might change if the number of Gibbs sampling iterations is high enough which we plan to investigate in the future.

Fig. 2 Two personalized outfits generated by GPT

In Fig. 2 we show two personalized examples generated by GPT. According to our fashion experts the first one fits perfectly and could have been created by a real stylist. The second one is acceptable, however, the color match between the cardigan and the coat could be improved.

5 Conclusions and Future Work

In this paper we have provided an experimental evaluation of Siamese Nets, Transformer, GPT, BERT, LSTM, and Seq-to-Seq LSTM on the outfit generation task using real customer data, both for personalized and non-personalized use-case. We have presented new adaptations on BERT, GPT, Siamese Nets and Seq-to-Seq LSTMs for this task and investigated how those have improved the model performance.

Within our extensive experimental results, we have confirmed that GPT outperforms BERT on outfit generation, while showing that BERT provides better performance on the FITB task. Moreover, we have compared personalized and non-personalized approaches, where we have showed that adding personalization does improve the performance of the algorithms with respect to expected customer engagement (e.g., CTR), which confirms that customers are not only looking for compatible outfits, but also for outfits that are of their taste. We have presented that the Transformer outperforms other models in terms of CTR, whereas Seq-to-Seq LSTMs provide higher personalization rate. We also shown that Siamese Networks are outperformed in both the personalized and non-personalized approaches.

As future work, we plan to investigate more sophisticated methods for personalizing BERT and GPT, such as allowing the models to attend to the personalization context with weights that differ from the self-attention weights instead of prepending it to the input sequence. Such changes have potential to lead to even better personalization. Moreover, we plan to extend our experimental results, while further improving them on the outfit generation task and providing A/B-test results for both GPT and the Transformer.

References

1. Bromley J, Guyon I, LeCun Y, Säckinger E, Shah R (1993) Signature verification using a siamese time delay neural network. In: Cowan JD, Tesauro G, Alspector J (eds) Advances in neural information processing systems 6, [7th NIPS Conference, Denver, Colorado, USA, 1993], pp. 737–744. Morgan Kaufmann (1993). http://papers.nips.cc/paper/769-signature-verification-using-a-siamese-time-delay-neural-network
2. Chen W, Huang P, Xu J, Guo X, Guo C, Sun F, Li C, Pfadler A, Zhao H, Zhao B (2019) POG: personalized outfit generation for fashion recommendation at alibaba ifashion. In: Teredesai A, V Kumar, Y Li, R Rosales, R Terzi, Karypis G (eds) Proceedings of the 25th ACM SIGKDD international conference on knowledge discovery & data mining, KDD 2019, Anchorage, AK, USA, August 4–8, 2019, pp 2662–2670. ACM (2019). https://doi.org/10.1145/3292500.3330652
3. Cheng W, Song S, Chen C, Hidayati SC, Liu J (2020) Fashion meets computer vision: a survey. CoRR abs/2003.13988 (2020). https://arxiv.org/abs/2003.13988
4. Devlin J, Chang M, Lee K, Toutanova K (2019) BERT: pre-training of deep bidirectional transformers for language understanding. In: Burstein J, Doran C, Solorio T (eds) Proceedings of the 2019 conference of the North American chapter of the association for computational linguistics: human language technologies, NAACL-HLT 2019, Minneapolis, MN, USA, June 2–7, 2019, Volume 1 (Long and Short Papers), pp. 4171–4186. Association for Computational Linguistics (2019). https://doi.org/10.18653/v1/n19-1423
5. Han X, Wu Z, Jiang Y, Davis LS (2017) Learning fashion compatibility with bidirectional lstms. In: Liu Q, Lienhart R, Wang H, Chen SH, Boll S, Chen YP, Friedland G, Li J, Yan S (eds) Proceedings of the 2017 ACM on multimedia conference, MM 2017, mountain view, CA, USA, October 23–27, 2017, pp 1078–1086. ACM (2017). https://doi.org/10.1145/3123266.3123394
6. Hochreiter S, Schmidhuber J (1997) Long short-term memory. Neural Comput 9(8):1735–1780. https://doi.org/10.1162/neco.1997.9.8.1735
7. Hu Y, Yi X, Davis LS (2015) Collaborative fashion recommendation: a functional tensor factorization approach. In: Zhou X, Smeaton AF, Tian Q, Bulterman DCA, Shen HT, Mayer-Patel K, Yan S (eds) Proceedings of the 23rd annual ACM conference on multimedia conference, MM '15, Brisbane, Australia, October 26–30, 2015, pp 129–138. ACM (2015). https://doi.org/10.1145/2733373.2806239
8. Koch G, Zemel R, Salakhutdinov R (2015) Siamese neural networks for one-shot image recognition. In: ICML deep learning workshop, vol 2. Lille (2015)
9. Kuhn T, Bourke S, Brinkmann L, Buchwald T, Digan C, Hache H, Jaeger S, Lehmann P, Maier O, Matting S, Okulovsky Y (2019) Supporting stylists by recommending fashion style. CoRR abs/1908.09493 (2019). http://arxiv.org/abs/1908.09493
10. Laenen K, Moens M (2019) Attention-based fusion for outfit recommendation. CoRR abs/1908.10585 (2019). http://arxiv.org/abs/1908.10585
11. Lee H, Seol J, Lee S (2017) Style2vec: Representation learning for fashion items from style sets. CoRR abs/1708.04014 (2017). http://arxiv.org/abs/1708.04014
12. Lipton ZC (2015) A critical review of recurrent neural networks for sequence learning. CoRR abs/1506.00019 (2015). http://arxiv.org/abs/1506.00019
13. McAuley JJ, Targett C, Shi Q, van den Hengel A (2015) Image-based recommendations on styles and substitutes. In: Baeza-Yates R, Lalmas M, Moffat A, Ribeiro-Neto BA (eds) Proceedings of the 38th international ACM SIGIR conference on research and development in information retrieval, Santiago, Chile, August 9–13, 2015, pp 43–52. ACM (2015). https://doi.org/10.1145/2766462.2767755
14. Mikolov T, Sutskever I, Chen K, Corrado GS, Dean J (2013) Distributed representations of words and phrases and their compositionality. In: Burges CJC, Bottou L, Ghahramani Z, Weinberger KQ (eds) Advances in neural information processing systems 26: 27th annual conference on neural information processing systems 2013. Proceedings of a meeting held December 5-8,

2013, Lake Tahoe, Nevada, United States, pp 3111–3119 (2013). http://papers.nips.cc/paper/5021-distributed-representations-of-words-and-phrases-and-their-compositionality

15. Polanía LF, Gupte S (2019) Learning fashion compatibility across apparel categories for outfit recommendation. In: 2019 IEEE international conference on image processing, ICIP 2019, Taipei, Taiwan, September 22–25, 2019, pp 4489–4493. IEEE (2019). https://doi.org/10.1109/ICIP.2019.8803587

16. Radford A, Narasimhan K, Salimans T, Sutskever I (2018) Improving language understanding by generative pre-training (2018)

17. Sutskever I, Vinyals O, Le QV (2014) Sequence to sequence learning with neural networks. In: Ghahramani Z, Welling M, Cortes C, Lawrence ND, Weinberger KQ (eds) Advances in neural information processing systems 27: annual conference on neural information processing systems 2014, December 8–13, 2014, Montreal, Quebec, Canada, pp 3104–3112 (2014). http://papers.nips.cc/paper/5346-sequence-to-sequence-learning-with-neural-networks

18. Vasileva MI, Plummer BA, Dusad K, Rajpal S, Kumar R, Forsyth DA (2018) Learning type-aware embeddings for fashion compatibility. In: Ferrari V, Hebert M, Sminchisescu C, Weiss Y (eds) Computer Vision - ECCV 2018 - 15th European conference, Munich, Germany, September 8–14, 2018, Proceedings, Part XVI, Lecture Notes in Computer Science, vol 11220, pp 405–421. Springer (2018). https://doi.org/10.1007/978-3-030-01270-0_24

19. Vaswani A, Shazeer N, Parmar N, Uszkoreit J, Jones L, Gomez AN, Kaiser L, Polosukhin I (2017) Attention is all you need. In: Guyon I, von Luxburg U, Bengio S, Wallach HM, Fergus R, Vishwanathan SVN, Garnett R (eds) Advances in neural information processing systems 30: annual conference on neural information processing systems 2017, 4–9 December 2017, Long Beach, CA, USA, pp 5998–6008 (2017). http://papers.nips.cc/paper/7181-attention-is-all-you-need

20. Veit A, Belongie SJ, Karaletsos T (2016) Disentangling nonlinear perceptual embeddings with multi-query triplet networks. CoRR abs/1603.07810 (2016). http://arxiv.org/abs/1603.07810

21. Veit A, Kovacs B, Bell S, McAuley JJ, Bala K, Belongie SJ (2015) Learning visual clothing style with heterogeneous dyadic co-occurrences. In: 2015 IEEE international conference on computer vision, ICCV 2015, Santiago, Chile, December 7–13, 2015, pp 4642–4650. IEEE Computer Society (2015). https://doi.org/10.1109/ICCV.2015.527

22. Wang A, Cho K (2019) BERT has a mouth, and it must speak: BERT as a markov random field language model. CoRR abs/1902.04094 (2019). http://arxiv.org/abs/1902.04094

23. Wu X, Lv S, Zang L, Han J, Hu S (2019) Conditional BERT contextual augmentation. In: Rodrigues JMF, Cardoso PJS, Monteiro JM, Lam R, Krzhizhanovskaya VV, Lees HM, Dongarra JJ, Sloot PMA (eds) Computational Science - ICCS 2019 - 19th international conference, Faro, Portugal, June 12–14, 2019, Proceedings, Part IV, Lecture Notes in Computer Science, vol 11539, pp 84–95. Springer (2019). https://doi.org/10.1007/978-3-030-22747-0_7

Understanding Professional Fashion Stylists' Outfit Recommendation Process: A Qualitative Study

Bolanle O. Dahunsi and Lucy E. Dunne

Abstract Unused and underutilized clothing is a major contributor to the environmental impact of the apparel industry. To reduce this underutilization, we need to implement ways to maximize clothing use. Artificially intelligent decision support may help users make better purchase decisions as well as daily dressing decisions. However, learning relationships between user and garment features is challenging due to the sparsity of data and the lack of validated expert models. One way to bridge this gap and inform clothing recommender system development is to understand how professional stylists choose outfits that maximize clothing use and satisfaction of clients. The purpose of this study was to understand how professional stylists make outfit and garment decisions for clients. This study used a qualitative approach to collect data from 12 professional stylists with varying areas of specialization on their decision-making process. Data were collected through semi-structured interviews and analyzed using thematic analysis. Findings show client features, garment features and the consultation process as the main factors in decision making. Consequently these factors could be integrated in design of recommender systems that increase consumers' clothing utilization.

CCS Concepts Information systems · Information retrieval · Retrieval tasks and goals · Recommender systems

Keywords Qualitative study · Requirements elicitation · Apparel recommenders

B. O. Dahunsi (✉) · L. E. Dunne
Department of Design, Housing, and Apparel, University of Minnesota, Twin-Cities, Minneapolis, MN 55455, USA
e-mail: dahun002@umn.edu

L. E. Dunne
e-mail: ldunne@umn.edu

N. Dokoohaki et al. (eds.), *Recommender Systems in Fashion and Retail*,
Lecture Notes in Electrical Engineering 734,
https://doi.org/10.1007/978-3-030-66103-8_8

139

1 Introduction

The global rate of clothing production has doubled from 2000 to 2014, with average consumer purchase increasing by about 60% annually and the clothes being kept half as long as they were 15 years ago [12]. Globally, the average annual loss of revenue due to underutilization and non-recycling of clothing is over USD 460 billion, with some garments getting just seven to ten wears [4], and as little as 7% of the wardrobe in regular use [3]. Most of these unutilized clothes get thrown out to make room for new ones and end up being reused, recycled, or thrown in landfills. According to the United States Environmental Protection agency, 11.15 million tons of textiles ended up in landfills in 2017, this constituted over 8% of total municipal solid waste for the year [17].

One way to reduce this waste is by improving consumers' clothing utilization. Consumers buy more clothes for multiple reasons. This could be because even though they have lots of clothes in their wardrobe they may not like those clothes, or they might not fit well. However, Dunne et al. found that a tiny fraction of even the clothing that consumers liked and thought was in regular use was actually used regularly [3]. This implies that consumers may not know how to wear what they have, so they buy more in hopes of finding better outfits. This then means even more complexity in the decision of what to wear and more underutilized clothes. Finding ways to improve the utilization of clothing that consumers already own could lead to reduction in their need to purchase more clothes as they would only purchase what they need. The purchases they make can then be fully utilized before being discarded. One solution to this problem would be to use recommender systems. Recommendations in most fields usually focus on making a single choice usually based on the user's preferences or based on how other users have rated items. In the case of recommending apparel, users may not trust their own preferences (they may want "expert" advice), and the preferences of others may not apply to their body type or aesthetics. The recommendation task is also complicated by the need to integrate each garment with other garments within the wardrobe system. Defining features for clothing recommendation (such as user features, context of use, and garment features that define clothing appropriateness) is a challenge that has not yet reached consensus.

Existing studies of apparel recommender systems each take a different focus. Some [8, 10] focus on finding matches for clothing using clothing attributes or images to identify tops or bottoms and find matches for them. While this method would be applicable for finding single complete outfits, it would not find pieces that integrate well with multiple items from a wardrobe. The other limitation of this method is that it does not match an outfit to the user's features it just matches it to other garments. Studies that account for context awareness also tend to focus on a sub-set of influencing factors. The study by [11] extends outfit matching by using context aware recommendations that considers weather input from the user or obtained from weather service websites to suggest outfits that match and that are suitable for the selected weather. Another context aware system by [9] uses the occasion to make recommendations. In their work, clothing attributes were used to classify different outfits and make recommendations based on how suitable they would be for different occasions. The main limitation of these context aware systems was that while they provide suggestions for individual outfits in the context provided, they do not consider the features of the user while making suggestions. The systems by [7, 16] on the other hand, integrate personal features in recommendations. These studies were on finding the right style for users given their body type, but didn't integrate other features like the context of wear or the user's wardrobe in suggestions. There have been a few studies on recommending from users' wardrobes. [6] designed a system that used RFID tagged clothing to implement a smart wardrobe with features of each garment integrated in it to track which outfits have been worn and where they were worn to. Although the system tracked what users wore and made recommendations from the wardrobe it did not use user features or contextual information for making recommendations. Although each of these recommenders focuses on different aspects, there is a need to design systems that put all of these features together for a more complex decision-making process. Although using recommender systems that integrate multiple features in purchase and styling recommendations could help simplify outfit choice and purchase decisions, it is important that the system uses enough features as well as the right combination of features to meet the user's needs. Each of these studies address crucial pieces of the recommendation task factors with slightly different variations in interpretation. Getting the perspectives of experts might be a way to standardize the inputs and models for each factor. This would help us determine what the right features are and how they should be applied in making recommendations that suggest good outfits for improved clothing utilization.

In order to design recommender systems that provide good outfit recommendations, we need to understand what good outfits are and what factors determine why an outfit is good for a user. A good source of information about what makes an outfit good in a given situation would be professional stylists who make such recommendations for clients daily. Understanding how professional stylists select outfits and make purchase decisions for clients could provide more insight into how to make recommendations for consumers on what outfits to wear or purchase. Although there seem to be numerous books, blogs, and media sources of information with style rules on how to style and what to wear, there is a lot of variation and contradiction in these sources. We would like to know how stylists navigate these conflicting or confusing advice and use them in making style decisions. The purpose of this study is to understand the factors that affect outfit choice from the perspective of professional stylists. These factors could then be used in designing recommender systems that act as automated personal stylists could help reduce waste from unutilized or underutilized clothing and reduce the impact of the fashion industry on the environment. The guiding research questions are:

What factors influence how professional stylists make decisions on outfit suitability and choice?

Do the outfit suitability and choice factors change if they are integrating the client's existing wardrobe or not?

To address the research questions, we conducted a qualitative study with professional stylists using semi-structured interviews. We sought to understand the factors they consider when making outfit choice for clients and their use and expectations from recommender systems.

2 Research Methodology

This was an exploratory study to understand the perceptions of professional stylists. The study was based on the interpretivist research paradigm of inquiry. The researcher's epistemological position for the study is that information is contained within the perceptions and experiences of professional stylists. Interaction with them, therefore, provides the opportunity to understand a complex problem with multiple themes. This study was conducted using phenomenology. Phenomenology is a qualitative research approach where you seek to understand the lived experience of participants [2]. The choice of phenomenology is appropriate here as it provides a good approach for comparison of participant responses to identify underlying themes. Semi-structured interviews were conducted to understand the decision-making process by finding out what experts think and why they make certain decisions in order to inform recommender systems design.

2.1 Interview Questions

A pilot interview was conducted with 3 colleagues to check for clarity, ambiguity, estimated length of interview and to minimize bias in the responses. The key interview questions were:

- What factors do you think are most important when choosing outfits for a client?
- Could you walk me through your typical consultation process with a first- time client?
- How do you decide what outfits to recommend to the client?
- Do you integrate the client's existing wardrobe in deciding what to recommend or just recommend new purchases?
- How do you integrate the client's existing wardrobe with new purchase recommendations?
- How would your process be different if you were integrating a client's existing wardrobe in your recommendation compared to if you are just making purchase decisions for a client?

Follow-up questions for the participants were asked to obtain clarification on issues they raised in their answers.

Table 1 Details of participants' area of specialization and years of experience

Participant ID	Specialization	Years	Location
QRI1	Personal shopping	10	Alabama
QRI2	Personal shopping	5	Minnesota
QRI3	Wardrobe consulting	17	Minnesota
QRI4	Author/image consulting	38	Minnesota
QRI5	Image consulting	10	Minnesota
QRI6	Personal shopping	7	Texas
QRI7	Personal shopping/wardrobe consulting	8	Minnesota
QRI8	Image Consulting	19	Louisiana
QRI9	Wardrobe consulting	6	Minnesota
QRI10	Personal shopping/editorial stylist	4	Minnesota
QRI11	Image consulting	4	Pennsylvania
QRI12	Image consulting	20	North Carolina

2.2 Participants

In this study, professional stylists were defined as anyone who had provided a minimum of 3 years of styling services to clients either as a personal stylist, wardrobe consultant, or image consultant. A total of 12 Participants were selected. The distribution of participants spanned the range of different job descriptions as shown in Table 1. Two of the participants were identified through referrals from colleagues. Four of the participants were identified by referrals from other participants and the remaining six were identified by a google search for the terms related to job descriptions in the professional styling field.

2.3 Data Collection

Each interview took between 30 min to one hour. Interviews were conducted virtually via Zoom or Skype video calls for 11 of the participants and in-person for one participant. All interviews were recorded. The in-person interview was conducted in a conference room on campus with the researcher and participant sitting across each other and a phone placed on the table between them to record the interview. The virtual interviews were recorded using the built-in software recorders and the phone voice recorder. Participants were either in their home or office during the video call. The COVID-19 social distancing restrictions and the geographic location of the participants during the interviews influenced the decision to interview most of the participants virtually. One advantage of the virtual interviews was that it allowed a more diverse set of participants to be included since geographic location was not a restriction. thus providing a richer source of data and information.

In order to focus on the participants during the interview notes were restricted to short insights or reminder words on questions that occurred to the researcher during the interview. The recordings were then played back within 30 min to one hour after the interviews. Notes were taken while listening to the recordings. The general impressions of the interview and context were also included in the notes.

2.4 Data Analysis

Interviews were transcribed verbatim and participant information anonymized using sequential codes assigned to each participant as shown in Table 1. The interview transcripts and notes were then imported into RQDA software for further analysis. The transcripts were read through for familiarity with the main concepts. Coding began after reading through all interview transcripts. At each stage of data analysis 'memoing' was used to reflect on impressions, decisions and thinking process behind the codes used. Memos also included observations on the data.

Thematic analysis was conducted using inductive coding for the data. The first phase of coding was done using open coding from the participants own words. Codes were grouped into categories and patterns identified. The transcripts were then recoded at a higher level to identify underlying meanings, processes, concepts, and salient points. At this stage some of the initial codes from the first round of coding were merged or redefined. The resulting codes were analyzed to find patterns and code categories were generated. The code categories were analyzed to find similarities and differences in responses of participants as well as possible explanations for those differences where differences were found. The generated themes were analyzed to answer the research questions.

3 Findings

Professional stylists that participated in this study had 3 major job descriptions. Personal shoppers took clients shopping or shopped for outfits and sent them to online clients. Wardrobe consultants helped clients reorganize the clothing in their closet as well as make new purchases to balance out missing pieces from their wardrobe. Image consultants worked with clients to build their personal or professional brand, so the client could portray the image they prefer to others. There were some who performed in more than one role and some who had other roles that were quite different from the three roles mentioned, but still related to fashion styling. Ten of the participants had both online and in-person clients, while two of the participants only had in-person clients.

From the analysis of the participant interviews three major themes were identified. These are discussed here with respect to the research questions and results of the data analysis. A separate theme on trust was also identified that was common to all participants and relevant to each of the research questions which is explained in a separate section after the research questions findings. Quotes from participants are integrated using the identification assigned to each participant as shown in Table 8.1.

3.1 What Factors Influence How Professional Stylists Make Decisions on Outfit Suitability and Choice?

The results of the data collected showed three major themes for factors that influence decisions of the stylists on outfit suitability. These themes were the client style considerations, garment related considerations and stylist consultation process. Each theme is explained in this section and a framework for the identified themes and the relationship between them is shown if Fig. 1.

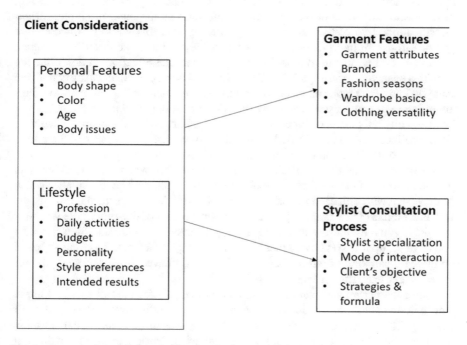

Fig. 1 Conceptual model of professional stylists' outfit decision factors

3.1.1 Client Considerations

Most of the responses in this theme were based on things that related directly to the client. Two subthemes emerged under this theme. They were personal features and client lifestyle.

Personal features: Participants responded that the personal features of the clients played a major role in outfit recommendations. Almost all the participants said they would first look at the body shape of the client in order to determine what would suit them. While all except one participant said body shape was important as it affected what style of clothing was appropriate for the client or if and where on the body, they could wear patterns, they were not specific about how the relationship works. This relationship will need to be investigated in more depth. Participants determined a client's body shape by visually assessing the body features of the clients during an in-person consultation or from the client's image, if it was provided, for an online client. Participants had different ways for classifying body types, but were consistent in using the difference in bust to waist to hip ratio as the main principle in their classification. Two of the participants specifically said that while it was common practice to use fruits to classify body shapes, they did not like describing women as fruits but rather just used those principles in their classification, "I don't like the 'you're a pear, you're an apple', I don't really like describing it that way. But it is true that you know, a person can be curvier on the bottom than on the top. And so things are going to fit that person differently than curvier on the top than on the bottom or being a really tall person or super petite."—QRI9. For participants who worked with clients online, they said they relied on the client to provide a description of their body features in order to try and determine their body type when they were dealing with clients online and couldn't visually assess the client. "So when I was working with people in person I can kind of just look and see like, Oh, they have a smaller waist, or they have wider hips, so this is going to work well. When it's virtual though, the way that I tell is [The Company] has another portion kind of like this where you have to input if you have narrow shoulders, average shoulders or wide shoulders and then you have to input if you have wider hips, narrow hips or average hips and so based on that you can kind of tell or they'll say like 'Oh, I have a longer torso or shorter torso, longer legs, shorter legs'."—QRI2.

Ten of the participants said they did a color analysis for their clients looking at their hair and skin tones to determine what colors of clothing works best with their coloring. The most common distinction they made was using coloring to determine if a client should wear warm or cool colors. Two of the participants said that the best colors to wear were affected more by the skin undertones of the client than the actual skin color of the client. "it has to do with your own color, your undertone. Okay, whether you're black or white. I have, and obviously I'm Black, I have clients that are Caucasian that are the same color palette as me."—QRI6. For participants who had tools they used for color analysis, their tools varied. Four said they used intuitive or trial-and-error methods, like draping fabric on the client to see which colors suited them. Three used strategies based on the color wheel while one participant said she uses a commercial color fan that has appropriate colors grouped on it based on the skin and hair color of the client.

While body shape and color were considered as important by most participants, other factors were seen as important by fewer participants. Three participants said age was important to them as outfits they recommended had to be age appropriate. The main consideration with age was about not recommending something skimpy or with inadequate coverage for an older client. Three of the participants said that they considered the client's personal perceptions of their body issues when making choices for them. They chose outfits that emphasized body features the client like while minimizing those features that the client wanted to hide.

Lifestyle: All of the participants had an initial consultation process where they either asked questions from the client, had them fill out forms that helped them figure out the lifestyle of the clients or used a pre-filled form from the client that was provided online. For most participants, the client's lifestyle played a major role in the choice of outfits. The lifestyle factors that were mentioned included the client's profession, daily activities, budget, culture, personality, style preferences and intended results. Almost all participants felt that profession and daily activities were important lifestyle factors to consider. Three of them said the profession of the client was important when purchasing or building a work wardrobe as different fields have different dress codes. Two of them said the daily activities of the client was very important as it determined what the client would be doing while wearing whatever outfits they chose. In some cases, the profession determined some of the client's daily activities. Participants said things like: "I need to understand who they are and what they're doing today and their life and where they want to go. When I know that, then I go in and assess their closet"—QRI5, "I ask a lot of questions about where they're from originally? What do they do outside of work? Do they have family, children, grandchildren? What hobbies they do, because I'm trying to get a mental snapshot of my client as much as possible so that I'm not placing or purchasing clothes on them that don't fit who they are, where they're trying to go"—QRI12, "like what type of Personality they have, what are they actually comfortable in, you know, how are they interacting? What is their lifestyle like?"—QRI3. The client's profession also might determine the daily activities of the clients in some cases.

Eight participants said they usually considered the personality or style preferences of the client, as it could determine what the client felt comfortable wearing or just preferred wearing generally. Although, they also addressed the need to balance what a client preferred with what the stylist feels the client should be wearing to help achieve their aims. Of those participants that felt personality was important, three said they would try to balance the client's personality with what they as professionals felt was most suitable for the client. For a client who has a more flashy personality but worked in an environment that has a more traditional dress code one of the participants said "I would say, why don't you get a scarf or handbag that will complement, you know, bring up your personality, but you still wear fairly standard suit and like a shirt, colored shirt, rather than a plain white shirt."—QRI11. Participants said they would sometimes need to remind the client that they hired them for a reason and try to explain to the client the reason for the choices.

Nine of the participants felt that the budget of the client was key. It determined where the stylist could shop for the client. Five of the participants said they usually started off shopping for most of their clients at second-hand clothing stores such as consignment or thrift stores. And then, depending on their budget would move on to departmental stores or boutiques. Culture was mentioned as an important factor by two of the participants. They spoke about the cultural requirements of their client being important in the outfit choices they make as different cultures had different expectations in terms of clothing. In some cases, it meant they had to integrate head coverings in the client's wardrobe or make outfit choices that required a lot more coverage than for most other clients.

3.1.2 Garment Features

Garment attributes: Analysis showed a theme of participant responses related to garment features. Some said they tried to find outfits for clients based on the garment attributes that were most appropriate for their personal features and lifestyle features identified. This meant trying to find the right fabric colors that worked for a person's skin undertones, or deciding where to put solids and patterns based on what body features they wanted to draw attention to. Some chose garments with fabrics or styles that would make them suitable for what the client was doing daily. This meant that they might recommend garments made with fabrics with a bit of stretch or a looser fit for comfort and ease of movement when the client's activities required a lot of movement, darker colors if the daily activities meant that clothing could get soiled easily or a corporate style with darker colors for someone in a formal job. One participant in choosing clothing for a stay-at-home mom who was an online client, discussed some things she would consider "So based on that and especially knowing that she's a stay at home mom, the things I'm going to prioritize again are like her hourglass figure for things that are going to compliment that. Things that you know nip in at the waist a little bit and then also she's staying home with her kids so comfort is probably really important to her, even if she is like going out, she probably wants something that could maybe also work for just staying home"—QRI2. She also mentioned some mistakes she made in her choices for the client "it looks nice on paper but isn't likely going to work well for her in her lifestyle. So a couple of goofs that I made was, the cardigan is super light gray, almost white and she has 3 kids. So, with three kids, it's easy to get messy and so she was like yeah, that doesn't really work for me. You know, it's too light of a color. She would have preferred a darker gray as a neutral." The client's profession also could affect what features of the garment made them appropriate for the client or what colors of clothing you need, "what do you have to do? Are you going to be speaking in front of people, do you have to do presentations for work? And do you have to address your boss about a situation, all those things have to be addressed…So I used to say if you're in a meeting with a bunch of guys the last thing you want to do is show up in a pink suit."—QRI5. In this case she said a suit in a darker shade would be most appropriate for a client who wanted to be taken more seriously.

Brands: Some responses in this theme related to things like how different brands were more suitable for certain client based on their lifestyle. For body shape they said the sizing and fit of some brands' cuts for things like jeans and pants may be more suitable for curvier women while some brands might have styles that are best for a petite or taller client. They said things like: "But the more you know about where to source and know about retailers and designers and how they fit and the cut, the better and easier it is for you to dress the client."—QRI12, "I think for instance I have one client, she loves Eileen Fisher because she's a shorter, heavier person and these clothes works for her and so whenever I get some of that brand and it is especially nice, I will call her"—QRI1. The budget of the client also determined if you shop at luxury brands, departmental stores or if you used second hand stores or not.: "Like understanding the range of what your client is willing to invest in, what they're willing to buy, and that drives a lot of my decisions of what brands to recommend when they're shopping for clothing, and what places I look at when I'm making decisions for them for styling."—QRI10. Some participants also said that different brands catered to different target markets, so some brands might focus on more formal clothing for work while others might specialize in more casual clothing.

Fashion industry seasons: Some participants said that since designers released new designs in store at different seasons, their choices depended a lot on what was available in stores during the consultation. The fashion season determined the fabric or colors that were in stores. So, for some clients, even if you knew certain colors would be best for them, those colors may not be in season at the time of the consultation, "when I talk about colors all their colors are not going to be in season... I mean, really, because there's certain times of the year, you're not going to get browns. Yeah, maybe Navy's aren't in all the time."—QRI5. The material, fabric and style of the garments also affected whether they were suitable for cold or warm weather. Some participants said they start by putting together client's wardrobe and outfit choices around the season that they are currently in and then help them select outfits for other seasons. "So that I'm only focusing on that kind of style or like the materials and if they're lightweight for summer or if it's for winter, if it's like heavier, mid-weight, kind of like pull over sweaters and stuff like that."—QRI10.

Wardrobe basics: For some participants garment features determined how suitable they were as wardrobe basics for clients based on their lifestyle or needs. This meant that they had to recommend pieces that were classics. "Classics are going to be stuff that does not go out of style. Your pinstripe suit, your A-line dress, your black pumps, stuff that don't go out of style. Or you could take a pencil skirt and mix it with anything. Yeah. That's the stuff that you're going to spend the most money on."—QRI8. For some the wardrobe essentials were independent of body type. Rather they depended on the industry the client worked in, "So it's not so much the essentials per body type, it's the essentials per industry. If I get a graphic designer or someone in the creative arts versus someone in banking, their essentials may be Chino pants and maybe a graphic T and a button-down. So it just really depends on the industry, more so than the body type." —QRI12. While for some the wardrobe basics are those garments that they felt every woman had to have in their wardrobe regardless or occupation or body type, "getting jeans that fit well because most people are wearing jeans 80% of the time. Getting one great pair of black slacks, whether it's a pair of black jeans if you're going to go that route or black dress pants. Getting some tank tops and shirts that skim the body, and getting a great blazer, like a jacket that fits well, and it curves in at the curves and that looks good. It's good at the shoulders. Those are some of the very, very basic things that I like to see all of my clients have. And that blazer can be any color, but it's just got to be a good fit and something you can use a lot" —QRI4, "I have what's called my ultimate wardrobe checklist, which is a list of garments that I feel every woman should have in her wardrobe for the most part and we go through and what's missing can also be the basis of a shopping list." —QRI6.

Clothing versatility: We found that based on the lifestyle factors of the client, some participants said they tried as much as possible to choose clothing that was versatile for clients. The versatility of the garments depended on how well the attributes of the garment translate to a different setting or occasion. Considering what the client was doing daily, they chose outfits that could be versatile enough to work in different situations with just slight modifications. This meant they might look at how the same garment might translate from day to night or from work to casual just by dressing it up differently with accessories, "so thinking about her lifestyle, I picked up pieces that would be comfortable that she could move around in but also pieces that are really versatile especially since she has that limited budget. So she can mix and match, she wants to be able to wear this during the day and then when she goes home and maybe wants to go out with a friend she can keep wearing the same thing or maybe just change the shoes and not have to put on a whole other outfit." —QRI2.

3.1.3   Stylist Consultation Process

Stylist specialization: We found that responses from participants were greatly affected by their area of specialization. This was most pronounced in the lifestyle factors that were emphasized by different participants. Image consultants seemed to emphasize the client's personal growth internally first then work on the outward appearance. For them they considered the style personality and the results the client wanted to achieve as being very important. Their emphasis was on the image the client was trying to project on the outside. They discussed working with clients on gaining confidence and building a brand and then designing the client's wardrobe around that brand. Personal shoppers felt that the occasion, season, body shape and lifestyle were more important. They also emphasized the need to help clients make purchases that were versatile and could be used for different events and occasions. The personal shoppers' emphasis on the versatility of clothing was understandable as most of them were not trying to work with a client's existing wardrobe, but rather helping clients to shop for new pieces for occasions or for their lifestyle. However, they said they might ask questions about what a client already owned or suggest other pieces that could be worn with what the client was purchasing. For wardrobe consultants, they usually were more concerned with things working together with existing pieces. All the wardrobe consultants said they shopped in the client's closet first by showing them what to keep or discard as well as how to wear what they already had differently. Then they would decide what new pieces needed to be purchased to improve the client's wardrobe. All stylists emphasized the need to have clients be comfortable in whatever they were choosing for them. Outfits needed to show those body parts the client wanted to emphasize and conceal those parts that the client wanted to hide. All the participants emphasized the need to teach clients and explain why they made the recommendations that they did.

Mode of interaction: The mode of client consultation was also important in the choices the participants made. This related to whether they were interacting with the client virtually or in-person. The main challenge with working with online clients for most participants was figuring out the color and body type of the client. For online clients they relied on information provided by clients in forms or on images sent by the client, and in cases where they were having a video conference then they could try to visually assess the client's shape. When it came to identifying an online client's coloring, they had to rely on whether the client described themselves as being dark or fair, as well as on what they could see of the client for online clients, whereas, for in-person clients they could visually analyze the client or do a color analysis using some of the tools mentioned previously. Participants also mentioned differences in how they approached online and in-person clients especially when making recommendations that integrate existing wardrobe. For online clients, they usually have accessed the client's previous purchases and the data the client provided from which they could build a list of likely pieces to recommend, and if this information was not available they worked mostly off whatever information the client provided in their forms about their body shape and lifestyle. For participants working with a client in-person, they tend to rely more on feedback from the client as they are working together pulling outfits and trying them on or if they are integrating the client's wardrobe they ask what they currently have in their wardrobe or look in the client's wardrobe to see what they have.

Client's objective: The objectives of the client in seeking a professional stylist's service was also a major factor for participants in outfit choice. They all said they tend to ask questions either in person or in the forms the client fills to understand the client and their main goals or objectives. "You also want to find out what is the reason for this change? They may be going through a divorce. They may have gone through cancer. So a lot of questions you want to ask to get to know the customer." —QRI8, "I think it's important that you take into consideration, who you're working with, that person and what their actual needs are, because everybody's needs are going to be different and you have to respect that." —QRI4. Some of the information they were trying to get also included: Do they want to find clothing for their current lifestyle or just want a total change of wardrobe? Are they trying to dress to achieve a certain result e.g. to get a bank loan, get a promotion, start a new career or improve their chances of getting a promotion? Are they dressing for a specific occasion, or just trying to get a new style? For some it's because they have a baby and need to find new clothing that works with the changes to their body. The client's motive also determines whether or not the wardrobe is integrated in the recommendations that the stylist makes. They emphasized the need to balance the client's motives, personality, or preferences with the expertise of the stylist. There could sometimes be conflict between what was better for the client in the opinion of the stylist and what the client believes was best. In such cases they sometimes had to remind the client of the reason why they consulted a stylist in the first place.

Strategies & Formulas: Some of the participants said they had to educate the client during the consultation on strategies they could use to transform an outfit. These strategies included showing clients how to mix up different pieces to get more wear out of the same pieces or how to style a garment differently for work or casual. A few said they showed clients how to build up mini wardrobe clusters around one piece by mixing it with other matching pieces. In some cases they would create look books for the client, so they had something to refer to later. It was important for them to show these strategies to the clients so they could do the same thing with the garments on their own later, "oftentimes it's just putting things together in different ways. And, you know, being more creative with how you put colors together. If their goal is to be more professional, showing how you can make an outfit more professional with accessories, how to use accessories, what kinds of accessories work better for professionalism and how jackets versus sweaters can change what an outfit looks like." —QRI9. In some cases the style and design of the garment may already be versatile. As one participant said about clothes she sent to an online client," And then for the dress I specifically picked this one because it ties at the waist. She's got an hourglass figure and its really going to flatter her, and then also it's reversible which you can't tell by the picture but you can actually flip it and it's like solid navy blue on the other side." —QRI2.

3.2 Do the Outfit Suitability and Choice Factors Change if They Are Integrating the Existing Wardrobe or not?

3.2.1 Existing Wardrobe Effects

Only one participant said that wardrobe integration did not change how she made decisions or what she recommended to clients, "On paper, a lot of people would think it does look very different. But honestly, what we're doing, I think, is the exact same because if I'm just shopping with someone, I'm still teaching them the things that I would teach in the closet, because I'm talking about styling and body type, and stuff like that the whole time and if I'm in a closet, I'm talking about the things that they need like for shopping and we follow up with the shopping list. So it's kind of the same, it's just one's in their home and one is out buying things." —QRI7

For all the other participants the factors considered in making outfit suitability decisions changed if they were integrating the existing wardrobe. The main difference was that when you were just choosing new outfits then you are looking at making recommendations for standalone outfits, but when you are choosing suitable outfits with wardrobe integration, you are making recommendations for a more complex case of new pieces needing to work well with already existing pieces. For standalone outfit recommendations they said they look mainly at the client's personal features like body type and color, then the lifestyle features such as the budget, the personality, occasion for which the outfit will be worn, they might also look at profession if the client is shopping for work wardrobe. "With someone who I'm styling some parts of their wardrobe that they already have, and I know about those pieces I can look for pieces that I can implement to mix in with those pieces. And rather than doing that with the other person… I try to focus on what occasion, we're looking for… And I try to figure out if the individual has staples that we can mix in with those, or if they don't, if I don't have that information, then I will try to mix in an array of like bottoms, tops or outerwear" —QRI10. When shopping with wardrobe integration on the other hand, they consider the factors that were considered for standalone outfits but may not consider the occasion as much. They would also need to focus more on the garment features. They look at things like versatility to see how well the new pieces work with existing pieces and what existing pieces can be paired with the new ones to change the look of the outfit.

Participants shopping with wardrobe integration would first shop in the client's wardrobe to see what is there and how those pieces can be combined to provide different looks then identify what is missing or what might be needed by the client before going shopping based on those needed items. Whereas when they are not integrating the wardrobe, they would go straight to shopping for new things. We found that although there were differences in the factors that determined suitability, most of the factors were still the same. Participants still tried to educate clients on why they choose some outfits for them by showing them how those outfits enhanced their body or color as well as showing them other ways they could use the same outfits. Some participants said they would still ask questions about what a client already owned when they were shopping without wardrobe integration as they wouldn't want a client purchasing something too similar to something they might already have.

Participants said they had more difficulty shopping with wardrobe integration when they worked with clients online. This was mainly because of the inability to really see the clothing of the client from an image or video and so they usually relied more on the past purchase history of the client when that was available. Still they said that this does not give a full understanding of the wardrobe, as they can only likely see what the client bought online and not everything they own. In those cases, they said they tended to rely on asking clients questions about what they have and may concentrate on purchasing garments that tend to work well with a lot of other garments and then make suggestions of ways the client might wear them.

4 Discussion

The purpose of this study was to understand how professional stylists make outfit choice and purchase decisions. In this section, we discuss two of the key themes identified: client considerations and garment features. We then discuss the third theme of the stylist consultation process and effects of wardrobe integration to explain the implications of this study for design of recommender systems.

4.1 Client Features

Personal features were a key consideration for participants but implementing these factors in a recommender system requires a lot more research. To design systems that could automate the styling process using recommender systems, body type and color would need to be implemented in the system. One way might be to require the users to fill in their body type but that would not be accurate, as has been shown in [14]. Their study showed that only about 45.1% of respondents' description of their body shape was consistent with their SizeUSA body shape classification. Given that the system would make predictions based on the body shape entered, the predictions would likely be wrong for more than half of the users based on this study. We found that in most cases the participants mostly relied on visual assessment to determine the body shape of the client, and when they couldn't see the client in person they would ask for descriptive details as well as full body images rather than asking the client what their body type is. Clearly, to successfully implement body type identification in recommender systems, a lot more study is required on ways to clearly identify body types as this cannot be left to the users to determine themselves.

While ten of the twelve participants said that color was an important factor, the approach, and tools for classifying a client's color differed. For example, three of the participants had a draping session to see which colors worked best with a client's skin tone, while one participant used color fans. There was also differences in what participants considered as most important for coloring. Two of the participants who discussed skin undertones said they felt the undertones of the skin color were more important than the actual skin tone while seven participants just looked at skin We can see from these findings that implementing color in recommenders, either as skin tone or undertones, is important. This could prove very complicated though, as in spite of recent advances in digital skin color identification, issues such as variation in screen representation of color and lack of standardized skin color names might make this difficult to implement. This clearly is an open area for future research.

Implementing some lifestyle features in recommenders would be easy. Budget can be easily implemented by setting limits on clothing recommended based on price or recommending from stores that provide clothing with the price range of the user. Other lifestyle features though could be more difficult to implement. Collecting accurate information on the user's physical attributes such as shape and color would require more research. Some lifestyle factors like personality, daily activities and profession might be inferred by having users answer carefully designed questionnaires that classify them but translating these inferred features into garment attributes would still be quite difficult. There is clearly a need for methods of digitally representing relationships between garment attributes and the lifestyle or user features they are suitable for. Although there are studies on using garment attributes in recommender systems [8–10], a lot more study needs to be done on how these attributes would translate in the kind of complex systems that would be needed to represent users existing wardrobe, as well as find new garments that could be added to the wardrobe to enhance it.

4.2 Garment Features

Understanding the relationship between different garment attributes and how they translate to lifestyle systems is essential in building these complex systems. One reason for this is that once the issue of how garment attributes and their relationship to different lifestyle features is addressed, the system can make recommendations using attribute filtering to find suitable outfits. One approach suggested from the literature to determine suitability could be using weighted scoring functions. This approach was used by [5] to find friend suggestions using weighted scores based on similarity of music interests. In the apparel recommendation case, weights could be assigned to garment features based on the suitability of those features for client considerations such as body shape, season, profession, etc. The total scores could then be used to rank outfits or garments to provide results. Attribute filtering could also be used to design and integrate versatility in the system to show how different garment pieces could be paired with others to create a different look. Garments with attributes that make them suitable for multiple situations could be used in determining the ideal set of wardrobe basics for each user based on their personal or lifestyle features. It could also be used for matching outfits using measures of similarity of features to find good pairs. There is also the added complexity of including diversity in the system so the user is not restricted to a narrow set of very similar outfit due to overfitting. The aim is to provide highly accurate recommendations based on user features while also including high diversity of styles in the recommendation.

One other important factor here is the difference in attribute feature names from person to person. While a stylist may understand the difference between a flared or pleated skirt, other users may not. The system needs to anticipate that users might either not know what different garment features are called or may have different names for different garment styles and be able to infer the style from the name and recognize garment attributes from images. Two possible approaches to this could be using natural language processing or image recognition. Natural language processing could be used to train the system on similarity in meanings between garment features while matching those names to garment images that could help users recognize what they mean so they can find similar items in stores. This would require a large dataset of style attribute names which is currently not available. While natural language could be used for handling name variation, the problem of image recognition is more difficult. From the literature we found studies using image recognition for determining garment attributes from images [1, 8]. These studies showed that the clarity of the images affected how well the system could recognize the garment attributes and even with clear images there were still limitations in performance of the systems. This would be especially important for trying to identify garments already owned by users from images.

4.3   *Implications for Recommender Systems Design*

All three identified themes are interrelated as shown in the conceptual model in Fig. 1. The client's features determine the garment features. A client's needs and lifestyle also affect which stylist they contact and the area of specialization of the stylist. For the stylist to be successful then, their decisions need to be based on the needs of their clients which affects what factors they prioritize in their decisions. To a large extent the specialization area of the participant affected which factors they felt were important. For example image consultants were more likely to look at profession and intended results in dressing as being more important while wardrobe consultants and personal shoppers emphasized more on daily activities and style preference as being important. Looking at the job function of each of these different fields, you find that the difference in emphasis by different specializations noticed in the lifestyle factors is understandable. A client would likely engage the services of a professional stylist that focuses on areas that are closely associated with the needs of the client. The stylist then places more emphasis on those factors that would increase the chances of successfully meeting the needs of their clients. This is also true for recommender system users. Users would likely have different intensions for using the system. Users looking to reorganize their wardrobe would likely require different features than those who just want to know what type of clothing works best for their body shape or color. Some might just want to find novel outfits that they wouldn't have thought of on their own, while others might just want to find new ways to wear what they already have. To provide the best recommendations for users there are different features that need to be integrated in the system design. The system needs to adaptable to what the user intends to find recommendations for. This would require modelling the system to elicit information from the user in order to learn what the user needs are and then adapt the recommendations to suit those needs.

We found from the study that stylists tend to provide recommendations that in most instances are quite different from what the client would typically get. This speaks to the need for serendipity and diversity in new outfit recommendations. Another peculiarity of the apparel industry is that style rules and seasonal trends change constantly. Recommender systems need to evolve with changing fashion rules as well as with the changes in fashion trends and seasons. There are existing studies on models that could anticipate and predict seasonal trends [18] which could provide the basis for designing the system to evolve in its recommendations. Another method that's been shown to work for this is by having the system learn new trends from recent magazines and social media posts of fashion influencers and integrate that to provide recommendations [19]. This has the added advantage of improving diversity and serendipitous recommendations.

In cases where the existing wardrobe affected the stylists' recommendations, stylist had to suggest outfits that were suitable for the client's needs and features, while also ensuring that new items can be used with existing items in the user's wardrobe. Recommendations that integrate a user's wardrobe could be designed with hybrid models where one part handles new outfit recommendations then another scores the suggestions based on how well they will integrate with the existing wardrobe. The outfits with the highest scores could then be returned in a ranked list for suitability. This has been used for recommendation in other fields for complex system but has not typically been used in apparel recommenders.

The stylists usually provided strategies and formulas for achieving a good look during the course of consultation. They also explained to users why they felt a certain outfit worked better for the client than another. This is important for recommender systems design as it emphasizes the need to provide explainable recommendations. This requires modelling the strategies and formulas used by professionals in the system to provide recommendations that are outside the user's normal choices while giving explanations for the recommendations. Using feature-based explanation might be preferable here as users can more easily relate to this and understand the explanations. This improves the trust and acceptability of the system and helps users better understand the choices made [15]. One approach to this could be using a rule-based algorithm that extracts user features and connect them to garment features from a knowledge base of professional expert rules based on a similar approach by [13] in their study. Although their study was done for connecting product features and user relationships, the same principles can be applied here using the user features and garment features with the knowledge base of expert rules as extraction rules to provide the explanation. The limitation of this method is that training the system to do this would require a large dataset of expert rules on body types and styles that suit them which is currently not available.

The aim of professional stylists and apparel recommender systems are the same. They need to find outfits that are ideal for the user. The only difference currently is in the level of personalization. The main difference is that professional stylists provide a more personalized experience that is tailored to the client's features and current needs while recommender system aim to do the same but in a way that scales to a larger audience so it may not be as personalized. Personalization of recommender systems for users can be improved using a threefold approach as shown in this paper. The first step is knowing which attributes of users to learn Then the system needs to learn the design-based attributes of clothing. The final step is determining what clothing attributes are suitable for which user attributes in order to make the right predictions based on expert provided rules.

This paper's main contribution is in helping to better understand which features to learn and provide suggestions on how to integrate the features in a recommender system. For some features, integration in a recommender system is a little more direct, e.g. collecting information on users' lifestyle and attributes. Requirements such as mapping of user attributes to garment features, translating context information such as profession and personality from user information into learnable models, developing datasets of expert rules for determining suitable outfits, etc. still requires more research before implementation of such complex systems can be possible. These areas provide open areas for future research.

5   Conclusions

This study was aimed at understanding how professional stylists make outfit choice and purchase decisions. Three main themes of client considerations, garment features and stylist consultation process were identified as the factors that affect how professional stylists make outfit choice decisions. Client considerations was found to have two sub-themes of personal features and lifestyle. The importance of some factors changed for most participants when existing wardrobe was integrated in the choice of outfits. While some of these factors have been implemented in existing systems, some problem areas identified provide areas for additional research in order for these factors to be fully implemented in recommender systems. Such systems would allow consumers make full use of their existing wardrobe while ensuring that new purchases are ideal for the users and integrate well with what the user already has in their wardrobe. Clearly, there are still a lot of areas that require extensive research before a fully context aware recommendation system of this can be implemented. There is also a need for research in well-defined expert styling rules and creation of rules dataset that could be used in model training for such systems.

The main limitation of this study is that the results are fundamentally the perceptions of the participants in the study, rather than the beliefs of professional stylists in general. The responses might not be generalizable as the participants are not a true representation of the professional stylist population. In spite of this limitation the findings could still provide a good basis for understanding the professional stylists' thinking process and guidance for implementing it in a recommender system.

Acknowledgements This work was supported by the US National Science Foundation under grant#1715200.

References

1. Chen JC, Liu CF (2017) Deep net architectures for visual-based clothing image recognition on large database. Soft Comput 21(11):2923–2939. doi:https://doi.org/10.1007/s00500-017-2585-8
2. Creswell JW (2014) Research design: qualitative, quantitative, and mixed methods approaches, 4th edn. SAGE Publications, Thousand Oaks
3. Dunne LE, Zhang J, Terveen L (2012) An investigation of contents and use of the home wardrobe. In: Proceedings of the 2012 ACM conference on ubiquitous computing (UbiComp'12). Association for Computing Machinery, Pittsburgh, Pennsylvania, pp 203–206. doi:https://doi.org/10.1145/2370216.2370247
4. Ellen MacArthur Foundation (2017) A new textiles economy: redesigning fashion's future. Retrieved Feb 18, 2020 from http://www.ellenmacarthurfoundation.org/publications
5. Fan C, Hao H, Leung CK, Sun LY, Tran J (2018) Social Network Mining for Recommendation of Friends Based on Music Interests. In: Proceedings of the 2018 IEEE/ACM international conference on advances in social networks analysis and mining (ASONAM), pp 833–840. doi:https://doi.org/10.1109/ASONAM.2018.8508262
6. Goh KN, Chen YY, Lin , ES (2011). Developing a smart wardrobe system. In: Proceedings of the 2011 IEEE consumer communications and networking conference (CCNC), pp 303–307. doi:https://doi.org/10.1109/CCNC.2011.5766478
7. Hidayati SC, Hsu CC, Chang YT, Hua KL, Fu J, Cheng WH (2018) What dress fits me best? fashion recommendation on the clothing style for personal body shape. In: Proceedings of the 26th ACM international conference on Multimedia (MM'18). Association for Computing Machinery, Seoul, Republic of Korea, pp 438–446. doi:https://doi.org/10.1145/3240508.324 0546
8. Lin Y, Ren P, Chen Z, Ren Z, Ma J, de Rijke M (2019) Improving Outfit Recommendation with Co-supervision of Fashion Generation. In: Proceedings of the The World Wide Web Conference on - WWW'19 (2019), pp 1095–1105. doi:https://doi.org/10.1145/3308558.3313614
9. Liu S, Feng J, Song Z, Zhang T, Lu H, Xu C, Yan S (2012) Hi, magic closet, tell me what to wear! 619–628. doi:https://doi.org/10.1145/2393347.2393433
10. Liu YJ, Gao YB, Bian LY, Wang WY, Li ZM (2018) How to wear beautifully? clothing pair recommendation. J Comput Sci Technol 33(3): 522–530. doi:https://doi.org/10.1007/s11390-018-1836-1
11. Liu Y, Gao Y, Feng S, Li Z (2017) Weather-to-garment: Weather-oriented clothing recommendation. In: Proceedings of the 2017 IEEE international conference on multimedia and expo (ICME):181–186. doi:https://doi.org/10.1109/ICME.2017.8019476
12. Remy N, Speelman E, Swartz S (2016) Style that's sustainable: a new fast-fashion formula. McKinsey&Company:1–6
13. Samih A, Adadi A, Berrada M (2019) Towards a knowledge based explainable recommender systems. In: Proceedings of the 4th International conference on big data and internet of things (BDIoT'19). Association for Computing Machinery, Rabat, Morocco, pp 1–5. doi:https://doi.org/10.1145/3372938.3372959
14. Song HK, Ashdown SP (2013) Female apparel consumers' understanding of body size and shape: relationship among body measurements, fit satisfaction, and body cathexis. Clothing Text Res J 31(3):143–156. doi:https://doi.org/10.1177/0887302X13493127
15. Tintarev Nava, Masthoff Judith (2015) Explaining recommendations: design and evaluation. Recommender Syst Handb 2015:353–382. https://doi.org/10.1007/978-1-4899-7637-6_10
16. Tu Q, Dong L (2010) An intelligent personalized fashion recommendation system. In: Proceedings of the 2010 International conference on communications, circuits and systems (ICCCAS), pp 479–485. doi:https://doi.org/10.1109/ICCCAS.2010.5581949
17. US EPA (2017) National overview: facts and figures on materials, wastes and recyclingǀ facts and figures about materials, waste and recycling US EPA. Retrieved 20 Feb 2020 from https://www.epa.gov/facts-and-figures-about-materials-waste-and-recycling/national-overview-facts-and-figures-materials#Landfilling

18. Yu Y, Hui CL, Choi TM (2012) An empirical study of intelligent expert systems on forecasting of fashion color trend. Expert Syst Appl 39(4):4383–4389. doi:https://doi.org/10.1016/j.eswa. 2011.09.153
19. Zhang Y, Caverlee J (2019) Instagrammers, fashionistas, and me: recurrent fashion recommendation with implicit visual influence. In: Proceedings of the 28th ACM international conference on information and knowledge management (CIKM'19). ACM, New York, NY, USA, pp 1583–1592. doi:https://doi.org/10.1145/3357384.3358042

Printed in the United States
by Baker & Taylor Publisher Services